UNION PACIFIC'S BIG BOYS

The complete story from history to restoration

Kalmbach Media

Kalmbach Media
21027 Crossroads Circle
Waukesha, Wisconsin 53186
www.KalmbachHobbyStore.com

The chapters in this book have appeared previously as articles in various issues of *Trains* magazine. Text, captions, drawings, and other information may have been edited and altered from their original presentations.

Published in 2020
27 26 25 24 23 4 5 6 7 8

Manufactured in China

ISBN: 978-1-62700-794-8
EISBN: 978-1-62700-793-1

Editor: Jeff Wilson
Book Design: Lisa Bergman

Library of Congress Control Number: 2020931281

On the cover: Newly restored Union Pacific Big Boy 4014 rounds a curve near Bucyrus, Kan., in November 2019. The iconic 4-8-8-4 is rolling on former Missouri Pacific rails en route to Kansas City Union Station. *Photo by Zach Pumphery*

CONTENTS

FOREWORD

by Jim Wrinn

GREEN RIVER, WYO., HAS BEEN A TOWN in the south-central part of this windswept state since before the railroad came. Almost 800 miles to the west from the starting point of the transcontinental railroad in Council Bluffs, Iowa, it came into being in 1868. It is a place right out of a cowboy movie with rocky buttes for a backdrop, and yes, the Green River does run through it. This town has seen it all when it comes to the Union Pacific Railroad. Its gaze has never flinched. Never missed a thing. Freight trains, passenger trains, work trains, wreck trains, executive trains. Helpers, hotboxes, hobos. Heroic engineers and firemen, vigilant brakemen and conductors. The eastbound procession of fruits and vegetables from California, lumber from the Pacific Northwest, potatoes from Idaho. The westbound passage of men and materiel for World War II, the Korean conflict, and the Vietnam War. Glamorous passenger trains with families to movie stars on board. The trains have literally come and gone millions of times in the last 150 years. The common denominator: Every train had at least one locomotive to propel it. For a short period in the mid-20th century, 25 of those locomotives were among the largest, most powerful, and greatest ever made: The Union Pacific Big Boy.

In the 1860s, the locomotives that pulled those trains were originally the classic American type, the 4-4-0 with their glittering and colorful dress, high-stepping drivers, and dainty pulling power. As the length and weight of trains grew, so did the size and power of the steam locomotives that pulled them. In the era before diesels took over in the 1950s, the pinnacle of steam-locomotive perfection was a brief, fleeting period of 18 years, starting just as the nation entered World War II. A single unit did the work of a dozen or more of those 4-4-0s and twice the work of the era's standard mainline freight and passenger locomotive, the 4-8-4 Northern. In less than 75 years, steam-locomotive technology evolved from one of the most basic designs to one of the most complex and most powerful. It was the era of Big Boy, among the largest and most powerful steam locomotives ever built in the U.S., North America, or anywhere in the world.

In these pages, you'll learn more about one of the most fascinating locomotives ever built — basically two powerful 4-8-4 locomotives under one massive 7,000-hp boiler. You'll learn why these locomotives were built, how and where they operated, and why they are no more. You'll learn why eight of the 25 — an amazing ratio for preservation — have been saved as monuments and museum pieces in their home territory as well as hundreds of miles away as far east as Pennsylvania and as far south as Texas. You'll learn why they were beloved: From their creation when an unknown shop employee chalked "Big Boy" on the smokebox at the factory where they were built to the thousands who ventured to the wilds of Wyoming and Utah to witness and record their passage and to those who kept the faith that maybe, *just maybe,* one might run again somewhere, sometime, someplace.

Big Boy is a classic component of the country's westward expansion. It's a classic saga of man and machine vs. gravity and tonnage. It's an American tale of creating the biggest and the best to solve a significant problem: wrestling freight trains across Utah's Wasatch grade,

Big Boy No. 4023 passes Hermosa, Wyo., during a climb up Sherman Hill with a 95-car eastbound freight, seen trailing into the distance around a wide S-curve in July 1950. *R.H. Kindig.*

5

a 65-mile-long sustained uphill climb between Ogden, Utah, and Evanston, Wyo. Their route was through scenic and dramatic Weber and Echo canyons, past natural landmarks like the Devil's Slide and Pulpit Rock. They saluted the sacred 1,000-mile tree, a fir designated by those building the transcontinental railroad. They clawed their way through the famous flyover at Curvo near the top of the hill. The Big Boy was built for this grade, but the fleet of locomotives also was put to the test on Sherman Hill, between Cheyenne and Laramie, Wyo., where Union Pacific crossed the Rockies at one of the lowest passes in the chain at some 8,000 feet. They showed their strength on Peru Hill, which started the moment the lead 36-inch guide wheels crossed the Green River bridge in its namesake town, and 68-inch drivers bit into 130-pound rail. And yes, it was built to sprint in the desert between Ogden and Cheyenne. Designed for speeds up to 70 mph, it most likely cruised at 50, but still a phenomenal feat for such a giant in an era when the average freight train speed was in the 20-to-30-mph range.

For those of us looking back almost 80 years, we can all be pleased that so many people traveled to the deserts of Utah and Wyoming to watch the daily spectacle of the Big Boy in action and at rest, in summer heat and in winter's fury. All of the great railroad photographers of the day made their pilgrimages to witness their majesty and might. Kindig. Perry. Hale. Shaughnessy. Then there was William Kratville, who turned his passion for all things Union Pacific into a landmark book (titled simply *Big Boy*) that is still the bible of the Big Boy to this day. UP understood the admiration the public had for the Big Boy — the company made a film about them called *The Last of the Giants.* All of these efforts to record and document the locomotives had to move fast — Big Boy's reign was less than a generation. And then they were gone. Save for the negatives, prints, a few videos, and books gathering dust on shelves, that should have been the end of the Big Boy story. A fading memory of a great locomotive. The stuff of history and legend.

But Americans are not satisfied with simply reading about their history. They want to experience it. They want to know what it feels like to stand beside the tracks as this monster thunders by. They want to hear the exhaust, the whine of the turbo generator, the melody of the whistle. They want to talk to the crew that handles this Industrial Age dinosaur. The story of Big Boy has an amazing ending with Union Pacific's decision in 2013 to restore one of the survivors. The UP steam team under Ed Dickens would scour the list of surviving engines and focus and settle on one far away from home, No. 4014, on display in Southern California. They would move the landlocked 600-ton beast more than a mile to reach an active track. They would spend months preparing the engine to return home to Cheyenne, then years planning, preparing, and executing the most significant steam locomotive restoration in the U.S. in more than 60 years. Only an organization as well-financed and well-equipped as the Union Pacific could successfully take on such a massive project. Restored in 2019 in time to celebrate the 150th anniversary of the transcontinental railroad, No. 4014 rolled west across Sherman and Peru hills and back across Wasatch grade in May 2019 to the delight of thousands of local residents and thousands more who traveled from across the world to witness this sight. And then it went on tour, going where Big Boys had never gone before. All of this was also a good reminder that naysayers are best left alone: It had been touted for years that a Big Boy would never be restored to operation because it was too big, too much to handle, too expensive to bring back to life. Thankfully, those who said such things were ever so wrong. And thankfully, Union Pacific management understands the company's role in U.S. history — its creation was authorized by Abraham Lincoln—and is proud to share it.

So not only can we study Big Boy's history and reincarnation in these pages, but we can examine the genuine article as it thunders down the main line. And that takes me back to Wyoming to a lonely hillside within sound of 18-wheelers on I-80. There, just west of Green River, Wyo., is a long sustained westbound climb known as Peru Hill. Today, if you scout the hillsides where dilapidated snow fences slowly decay next to the curving tracks, you can find evidence that giants once roamed here. In the dirt and sand that form the hillsides are deep deposits of cinders placed here when Big Boys came thundering upgrade with long strings of freight cars on their drawbars. They're here to give silent testimony to the legend that once was the Big Boy. If you close your eyes and let your mind wander across the Green River, you can travel back to a day when the Big Boys called their railroad their own, and for good reason. They were the undisputed Kings of Steam! They always have been. They always will be.

Jim Wrinn
Editor, Trains magazine
April 2020

At the crest of the Continental Divide, Big Boy 4-8-8-4 No. 4023 charges east past Challenger 4-6-6-4 No. 3965 westbound on Oct. 22, 1956, after both pounded up a 0.82 percent grade to 7,107 feet above sea level at Creston, Wyo. *Alex Bremner Jr.*

1 Birth of a legend

by Jim Wrinn

When Union Pacific's 4-8-8-4s ruled the west

NOW THAT UNION PACIFIC'S RESTORATION of Big Boy No. 4014 is complete and the 4-8-8-4 is active again, admirers of this huge machine would do well to remember two names: Otto Jabelmann and William Jeffers. You could consider them the fathers of this amazing, unique locomotive that was one of the most successful articulated steam locomotives in North America.

The story of the Big Boy locomotive is a tale of human ingenuity and innovation, rapidly changing technology, and the eternal enchantment with overwhelming power.

Jeffers, as Union Pacific's president in 1940, instructed Jabelmann's Research and Mechanical Standards Department to design and build larger motive power to conquer the grades of the Wasatch Mountain range east of Ogden, Utah. Jabelmann was the perfect man for the task, as he had been instrumental in the creation of the company's high-speed passenger 4-8-4s and its highly successful multipurpose 4-6-6-4 Challengers.

While the majority of Union Pacific's Overland Route from Omaha, Neb., to the Southern Pacific interchange in Ogden was populated with relatively easy grades, the Wasatch posed a significant barrier: The eastbound run of 65 miles from Ogden to the summit at Wahsatch, Utah, was beset by grades of 1.14 percent. Since the opening of the transcontinental railroad in 1869, UP had designed larger and larger steam power to conquer the Wasatch range. The Big Boy was the culmination of those designs.

According to the seminal 1963 book *Big Boy* by the late William W. Kratville, the Research and Mechanical Standards Department was established in 1936 under Vice President Jabelmann. Jeffers' order to Jabelmann in 1940 was to develop a locomotive capable of pulling 3,600 tons over the Wasatch unassisted. To do so, Kratville recounts, the locomotive would need 135,000 pounds of tractive effort and an adhesion factor of four. Mechanical engineers concluded that to meet Jeffers' demands would require an articulated with two sets of eight driving wheels, four guide wheels, and a four-wheel trailing truck to support the enormous furnace necessary to supply steam. Thus the 4-8-8-4 wheel arrangement was born.

Within three months, a design team had formed. The UP had long been associated with the American Locomotive Co., and the railroad and the builder collaborated in the creation of the Big Boy. Because UP had accumulated significant research data, the entire project took only about a year to complete: six months to design, fabricate, and acquire parts, and another six months to construct the first locomotive.

Union Pacific's confidence in the initial design was evident with its order for 20 Big Boys from Alco at a cost of $265,174 each, or about $4.2 million in today's dollars.

The design was nothing short of an operating man's dream come true: built to run at speeds of up to 80 mph, take a maximum curvature of 20 degrees, and produce maximum power at 30-40 mph. In reality, they rarely got to run fast, but the overdesign meant the engines could put on a spectacular performance, day in and day out.

While the new engines were being built, Union Pacific worked to prepare the railroad's infrastructure for these colossal machines, which weighed 600 tons in operating order and spanned nearly 133 feet in length with a wheelbase of 117 feet, 7 inches. The railroad rebuilt bridges, realigned curves, and laid heavier 130-pound rail between Ogden and the summit

Extra 4014 West during the steam era: The 4-8-8-4 pulls a merchandise train toward the summit of Sherman Hill in the early 1950s.
James L. Ehernberger

Long before it became a star, Big Boy No. 4014 was just the 15th engine of an initial order for 20 Union Pacific 4-8-8-4s completed in fall 1941 at Alco's plant in Schenectady, N.Y. On the eve of World War II, the engines were pressed into service. *Emil Albrecht; James L. Ehernberger collection*

THE BIG BOY story

1940

UP President William Jeffers asks his research and mechanical standards department for bigger power to tackle grades and tonnage in the Wasatch Mountains.

1941

SUMMER

Railroad eases curves, installs heavy rail to accommodate the new power.

SEPTEMBER

- American Locomotive Co. completes Big Boy No. 4000, the first of 20 such locomotives.
- The locomotives generate more than 500 news stories while en route via Delaware & Hudson, New York Central, and Chicago & North Western.

James L. Ehernberger collection

NOVEMBER

Alco finishes No. 4014.

at Wahsatch. Because of the great overall length of Big Boy and the swing of both boiler and cab, the UP determined that sections of double track were placed too close for a 4000 to clear passing trains. Where curves could not be eased, the railroad instituted passing and meet restrictions to preclude damage to the locomotives or wide loads. Track work began as the railroad identified problem curves and adjusted track centers to provide more clearance.

One of the most critical concerns was turning the giant locomotives at endpoints. To do this, UP installed 135-foot turntables at the key servicing points of Ogden; Green River, Wyo.; and Laramie, Wyo., in late 1941. Cheyenne, being more space constrained, received a 126-foot table. Denver had a suitable wye track that would allow turns. A 126-foot turntable, installed at Caliente, Nev., in 1941, was moved to North Platte, Neb., in 1950 to allow easier turning of the big articulateds.

While the initial Big Boy haunt was between Ogden and Green River, as World War II progressed, the Big Boys' official operational territory extended east to Rawlins, Laramie, and Cheyenne. From time to time, they also visited the giant yard in North Platte. They were cleared to operate between Salt Lake City and Pocatello, Idaho, and between Salt Lake City and Los Angeles, although it has not been documented that they ever reached those far-flung locations.

Naming the Big Boy

The name "Big Boy" came about by accident. An unknown Alco employee in Schenectady, N.Y., chalked the name "Big Boy" on No. 4000's smokebox while that first 4-8-8-4 was under construction. The name, appropriate and catchy, stuck, although it was rumored that UP had considered naming the class "Wasatch." (TRAINS, incidentally, disdained the Big Boy name at the time, suggesting in 1943 that if UP's 4-6-6-4 high-speed articulateds were known as Challengers, then it would only be right to call the larger 4-8-8-4 class Champ, short for champion, "for champ it is, in horsepower and size.")

Number 4000 was shipped dead via Delaware & Hudson, New York Central, and Chicago & North Western to Council Bluffs, Iowa. A UP switch engine towed the Big Boy across the Missouri River to Omaha Shops where it was officially accepted on Sept. 5, 1941. The 4000 was steamed up, then put on display at Omaha Union Station. A few days later, it traveled light to Council Bluffs for servicing, then back to Omaha to pick up a train of 100 empty Pacific Fruit Express reefers. The locomotive made several stops as it traveled west across Nebraska for water, fuel, and crews. It left Cheyenne going west on September 8.

The Big Boy class arrived on the eve of World War II, and the ensuing crush of traffic meant the locomotives were extremely busy in their first years of service.

The first assignments for the 4000s were on the Wasatch grade in Utah, and the locomotives immediately improved operations. Working between Ogden and Green River, the 4000s allowed Union Pacific to reassign Challengers east of Green River where their increased tractive effort was badly needed to handle wartime traffic. East of Green River, the early Challengers (Nos. 3900-3939) could handle the same tonnage as the Big Boys on the Wasatch. As a result, UP could begin to operate priority green fruit blocks with minimal switching moves and reduced terminal delay.

TRAINS collection

1942

DECEMBER

No. 4014 arrives on UP property.

Big Boys muscle wartime traffic across the U.S., averaging 7,000 miles per month.

1943

Big Boys No. 4004, 4014, and 4016 are tested against a three-unit set of diesel locomotives.

1944

Alco builds five additional Big Boys, Nos. 4020-4024.

1959

JULY 20 & 21, 1959

- No. 4014 completes its final revenue freight run from Laramie, Wyo., to Cheyenne, Wyo., a trip of 3 hours and 35 minutes, beginning at 10:15 p.m. on July 20 and tying up at 1:50 a.m. the next day.
- No. 4015 makes the final run of a Big Boy, arriving in Cheyenne at 7:55 p.m.

TRAINS collection

The first Big Boy, No. 4000, makes its initial trip from Ogden, Utah, to Green River, Wyo., in 1941, flying the white flags of an extra (the "X4000" in the train indicator boards means this is train is Extra 4000 East). The car tucked behind the tender allows the mechanical department to monitor the 4-8-8-4's performance. *E.C. Schmidt*

1961

OCT. 18, 1961

Railway & Locomotive Historical Society's Southern California Chapter sends letter requesting donation of a Big Boy steam locomotive to UP President Arthur Stoddard.

James L. Ehernberger

DEC. 7, 1961

UP replies that it will donate No. 4014.

DEC. 20, 1961

Chapter sends acceptance telegram to UP via Western Union.

DEC. 29, 1961

After winter storms clear, No. 4014 is painted and prepared for the journey to Southern California and departs Cheyenne in tow at speeds up to 25 mph.

1962

JAN. 3, 1962

Train transporting No. 4014 arrives in Las Vegas, Nev.

Its initial career is just two years from its conclusion — and its revival by Union Pacific is still more than six decades in the future — as Big Boy No. 4014 heads west across Sherman Hill at Emkay, Wyo., in summer 1957. *James L. Ehernberger*

1985

Near San Bernardino, Calif., on Jan. 6, 1962. John E. Shaw Jr.

JAN. 5, 1962

No. 4014's train moves from Las Vegas to Yermo, Calif.

Near San Bernardino, Calif., on Jan. 6, 1962. John E. Shaw Jr.

JAN. 6, 1962

No. 4014 leaves Yermo, moves to Southern Pacific at Colton, Calif., and arrives at Bassett.

JAN. 8, 1962

Pacific Electric moves Big Boy No. 4014 to its new home at the Los Angeles County Fairgrounds in Pomona, Calif.

1985

RailGiants Museum moves within the Los Angeles County Fairgrounds.

Decoding the cab-side data

Like many railroads, UP affixed all of the basics of each steam locomotive in a code spelled out under the road number on the cab sides. *Jim Wrinn*

4-8-8-4 is the Whyte classification wheel arrangement: A four-wheel pilot truck, followed by eight driving wheels, followed by another eight driving wheels, and a four-wheel trailing truck.

1 following the wheel arrangement indicates the first subclass of that wheel arrangement.

68 is the outside diameter in inches of new driving wheel tires, excluding the flange.

23¾ 23¾ above the fraction bar indicates the inside diameter of the front and rear cylinders; 32 below the bar is their piston stroke length, all in inches.

540 is the total weight on drivers in thousands of pounds (540,000 pounds). MB identifies the stoker as a modified type B. — *Gordon McCulloh, UP steam historian*

Big impact on operations

Delivery of the 4000s affected operations elsewhere on the UP system. As Big Boys were assigned to the Wasatch, they dramatically reduced the use of helper engines. Wyoming Division's 9000-class 4-12-2s moved east to the Nebraska Division, displacing 5000-class 2-10-2s. These locomotives in turn went where they were most needed, some as helpers and others displacing Mikados, and so on down the ladder.

But it was not long before the Big Boys were put to the test to determine their own fate. Number 4014 was among three locomotives selected for this experiment. Along with Nos. 4004 and 4016, No. 4014 was involved in a test against a three-unit diesel in April 1943 between Ogden and Evanston. According to Kratville's book, on April 2, 1943, No. 4014 took 65 cars totaling 3,479 tons out of Ogden. All the way upgrade, the throttle was open less than full, and yet the 4014 accelerated at points on the grade from 1.8 to 4.5 mph per minute. A top speed of 42 mph was recorded on level track, while the minimum speed was 13 mph on a 3-degree curve on a 1.14 percent grade. Following tests with the other two Big Boys and the diesels, the internal combustion power proved to do no better than the steam engines, and the railroad concluded that steam would remain on the route.

Sadly, Jabelmann never saw this magnificent three from his greatest creation put to the ultimate test: He died Jan. 6, 1943, while in England on a special mission connected to Lend-Lease distribution, the program under which the U.S. government provided American industrial goods for the defense of allied nations.

2012

DECEMBER
Union Pacific publicly indicates its interest in acquiring and restoring a Big Boy.

2013

JUNE
Southern California Chapter of Railway & Locomotive Historical Society votes 8-2 to transfer No. 4014 to Union Pacific in exchange for a diesel and caboose to be displayed at RailGiants Museum, paving the way for the Big Boy's restoration.

AUGUST
UP steam crew arrives at museum to begin preparations for move.

NOVEMBER
Process of inching No. 4014 out of the fairgrounds begins.

Jim Wrinn

NOV. 15-DEC. 26
Moving process continues, bringing No. 4014 closer to Metrolink tracks it will use to leave fairgrounds.

2014

JAN. 26
UP's Southern Paci[fic] heritage unit, No. 19[…] pulls No. 4014 out o[f] the fairgrounds.

Twin pillars of smoke mark the double exhaust stacks that were adopted as standard for Union Pacific freight power. Hidden are the many changes inside the elongated smokebox that over the years were modified to improve drafting and reduce the chance for lineside fires. *Trains collection*

Thomas Bunce

FEB. 1-2, 8-9

No. 4014 is put on public display at UP's West Colton yard.

APRIL 28-MAY 8

With 11 days of moves and public displays, the Big Boy is taken from California to Cheyenne, Wyo., for restoration at UP's steam shop.

Jim Wrinn

Big Boy had won a big fight, but the motive power competition of the 20th century — steam vs. diesel — raged on, and in fact, was just beginning.

As World War II continued in 1944, trains of cargo and troops crisscrossed the country, and the first set of Big Boys was due for heavy shopping. Union Pacific received authority from the War Production Board to build five more Big Boys, Nos. 4020-4024, at a cost of $319,600 each. They were identical to the first set except for the use of heavier metals in the boilers and rods. The arrival of the second set meant the Boy Boy's territory was extended from Green River to Cheyenne.

Earlier designs benefit the Big Boy

As UP's largest steam power, the Big Boys benefited from years of improvement, and testing with the road's smaller power contributed to the success of the locomotive. As a result, the 4000s shared design features with UP's other modern steam locomotives.

Cast-steel frames with integral cylinders had first been used on Union Pacific in 1930 with the final order of 4-12-2s, and they proved superior to the earlier built-up frames. However, when the Challengers were ordered in 1936, the railroad went back to built-up frames to save on their initial cost. By the time the 4-8-4s were ordered in 1937, the company had decided that cast-steel frames were superior to the built-up type, and the cast-steel strategy was adopted as standard, along with roller bearings on all axles. Sixteen General Steel Castings Corp. Boxpok (pronounced "box spoke") disc driving wheels would assure a smooth ride at any speed for the 4-8-8-4s.

The Big Boys' nickel-steel articulating side rods had mating tongue and jaw ends that rode on bronze-lined mild steel bushings, which in turn pivoted on the driver crank pins. These were designed for and first used on the 4-8-4s, after which they were designated standard for new power. By eliminating knuckle joints, rod mass was significantly reduced, thus greatly simplifying the balancing process. The Big Boys used four Nathan DV-7 lubricators for their moving parts. The original lever operation of the front two lubricators was converted to a chain-drive arrangement by 1952.

The Standard Stoker Co.'s Modified Type B (or MB) stoker was specified standard for freight power with the 4000s because of its ability to reliably deliver more coal than even a Big Boy might consume. Jabelmann-patented Labyrinth front ends were adopted as standard for new power in 1939, being used on the second batch of 4-8-4s and the Big Boys. About 1943, UP modified the front ends, but this was replaced by an Improved Master Mechanic's design by 1946 to reduce the number of trackside fires caused by sparks emitted from the twin stacks.

The double exhaust stacks were adopted as standard for freight power with Big Boy as new perspectives were emerging to improve the drafting and efficiency of large power. Cast-steel pilots first seen on 4-8-4s were upgraded with a swing coupler, being much easier to operate than the drop-coupler version. Pilot-mounted air cooling radiators on the first 4000s, the maze of coils on either side of the headlight, were put behind the air pump shields by 1951.

Another strategy introduced in 1941 was to mount the turbogenerator close to the

2016

AUGUST

After concluding work on 4-8-4 No. 844, restoration work begins in earnest on No. 4014.

2019

FEB. 6

No. 4014 passes its hydrostatic test.

MAY 2

With restoration work complete, No. 4014 debuts with a test run from Cheyenne to Nunn, Colo.

MAY 9

No. 4014 joins No. 844 as part of a celebration of the 150th anniversary of the Golden Spike in Ogden, Utah.

Jim Wrinn

Union Pacific continued to look for ways to improve the Big Boys throughout the locomotives' lifespan. Number 4019, leaving Ogden on Jan. 19, 1946, sports "wind wings," or smoke deflectors. *Emil Albrecht; James L. Ehernberger collection*

4-8-8-4 Facts

Builder	American Locomotive Works, Schenectady, N.Y.
Date	Nos. 4000-4019, 1941 Nos. 4020-4024, 1944
Cost when new	1941: $265,174 each; 1944: $319,600 each ($4.2 million in 2014 dollars)
Length over couplers	132 feet 9⅞ inches
Weight in working order	First 20: 1,189,500 pounds (762,000 engine, 427,500 pounds tender) Last five: 1,208,750 pounds (772,250 engine, 436,500 tender)
Tractive effort	135,375 pounds
Horsepower	Approximately 7,000
Factor of adhesion*	4.02 / 4.03*
Fuel	Soft coal
Boiler pressure	300 pounds per square inch
Top speed	80 mph

Front guide wheels and trailing truck under firebox	Front: Inside roller bearing, 36 inches diameter Trailing: Outside roller bearing, 42 inches diameter
Driving wheels	68-inch Boxpok type with roller bearings and heat indicators
Wheelbase	18 feet 3 inches on both sets of drivers; 72 feet 5½ inches on engine; 117 feet 7 inches, engine and tender
Running gear	Driving axles, main and side rods, crankpins and piston rods made of heat-treated, low-carbon nickel steel
Pistons	Lightweight alloy steel; combination bronze and cast-iron piston packing rings
Cylinders	23¾-inch diameter by 32-inch stroke
Crossheads	Manganese-vanadium alloy steel castings in multiple-bearing guides made of carbon-steel forgings

Weight of reciprocating parts on each side of the locomotive	2,106 pounds on the front engine; 1,912 pounds on the rear engine
Steam distribution	12-inch piston valves controlled by Walschaerts valve gear
Lubrication	Four 36-pint mechanical lubricators with a total of 49 feeds leading through two- and four-way dividers and terminal checks to 123 oil outlets that oil the valves, cylinders, piston rod packing, steam pipe and exhaust pipe ball and slip joints, driving boxes and wedges, guides, engine-truck and trailing-truck center plates, trailing-truck journal boxes, engine-truck lateral motion device, throttle, reverse gear, articulation hinge pin, stoker, furnace bearers, and radial buffer
Grease fittings	Spring-rigging joints, brake-rigging pins, and valve-motion bearings
Boiler	Three courses fabricated from cold-rolled and stress-relieved steel
First course	Conical with a front inside diameter of 95 inches; 1 11/32 inches thick
Second course	1 3/8 inches thick
Third course	106 9/16 outside diameter; 1 3/8 inches thick
Firebox	235 1/32 inches in length by 96 3/16 inches in width, 150.3 square feet
Superheating surface*	2,043 square feet / 2,466 square feet*
Evaporating surfaces in square feet*	Tubes: 967/2,734. Flues: 4,218/2,301. Firebox: 593/595. Circulators: 111/125. Total: 5,889/5,755*

Tubes*	75/212 2 1/4-inch diameter, 184/73 4/5 1/2 inch diameter, all 22 feet long*
Combustion chamber	112 inches long
Crown sheet	27 feet long
Smokebox	Three-piece welded design
Flexible staybolts	MK-type caps
Firebox	Six sliding-shoe furnace bearings. Eight security circulations
Boiler	Foam-collapsing trough and manual blow-down system
Stoker	Standard modified type B
Boiler feed water equipment	Live-steam injector on the right side; exhaust-steam injector with remote control and centrifugal pump on the left side; twin stacks; exhaust nozzle tips consisting of four jets; modified Master Mechanic front-end netting
Cab	Supported from the boiler, insulated with Fiberglas and lined with Masonite. Air defrosters and windshield wings; seating for engineer, fireman, road foreman of engines, and brakeman. Two water columns with an upper glass for running on level or ascending grades and a lower glass for descending grades. Two saturated steam turrets supplied via 3 1/2-inch dry pipe from dome. Third superheated steam turret supplies stoker, turbo.
Pilot	Integral casting with front bumper beam. Swing-type coupler that folds into housing
Brakes	No. 8 ET
Air compressors	Two 8 1/2-inch cross-compound. 15 feet of radiator pipe between air pumps and sump reservoir
Tender*	Water bottom bed of the 4-10-0 wheel arrangement. 42-inch diameter wheels with roller bearings. Coal capacity: 28 tons. Water capacity: 24,000 gallons/25,000 gallons*

*Figures are for Nos. 4000-4019 / Nos. 4020-4024, respectively. Photo: Jim Wrinn

The seven-axle Centipede tenders of the Big Boys held 24,000 gallons of water and 28 tons of coal. The tender with No. 4014, shown as it is taken to Pomona, Calif., for display, was originally delivered with locomotive No. 4015. *Trains collection*

Number 4021, one of five locomotives from UP's 1944 order of a second group of Big Boys, prepares to lead a westbound extra out of Cheyenne toward Sherman Hill on June 24, 1956. Their construction required War Production Board approval. Use of heavier metals meant these Big Boys weighed 19,000 pounds more than the originals. *Jim Ehernberger; Al Chione collection*

roadbed. These were placed on the right front of the trailing truck on Big Boy. However, because of excessive contamination, they were moved to the boiler top starting in 1946, greatly improving their reliability.

Use of a single lateral cushioning, or lateral motion device, was first seen on large four- and five-axle locomotives. Two were used with the Union Pacific's 4-12-2s, with one at each end of the driver set.

On UP's 4-8-4s, with a driver wheelbase of 21 feet, 6 inches, and a need to operate at up to 100 mph, the mechanical department decided to employ lateral motion devices on axles 1 and 3 to ease curving. That strategy worked well. Thus as the second batch of 4-8-4s evolved, to further ease their huge drivers at high speeds, another lateral motion device was added, so that all except the last driving boxes were allowed lateral movement. As the Big Boys' design moved forward, this principle became standard for new power, thus this engine would, like the 4-8-4, have no rigid wheelbase. Later power was similarly configured.

A tender story

The Big Boy also adopted the seven-axle Centipede tender introduced with Union Pacific's second batch of 4-8-4s. The Centipede design allowed mechanical engineers to limit the tender's length to barely more than 47 feet, while in time allowing some increase in capacity for both fuel and water. All UP Centipede tenders were the same length, being just 15 inches longer than the 20,000-gallon, six-axle types the first batch of 4-8-4s sported. Cast-steel water bottom frames had been the UP standard for new tenders since the first of the big 18,000-gallon cylindrical tenders arrived with the 4-12-2s in 1928.

The first 20 Big Boys were ordered with 24,000-gallon Centipede tenders that held 28 tons of coal. As the first group of late Challengers was ordered in 1942, their tenders were similar to the Big Boys' tenders, but by altering their fuel space the designers increased water capacity to 25,000 gallons.

UNION
PACIFIC
4014

With a clear stack, an exhaust arching nicely in the Wyoming sky, and a tender full of coal, No. 4014 departs Cheyenne going westbound in the 1950s. *Emil Albrecht; James L. Ehernberger collection*

The end is at hand for the original Big Boy, as the shield number plate is gone and workers remove No. 4000's bell at Cheyenne in 1961 ahead of the locomotive's scrapping. Eight of the 25 Big Boys were saved from the scrapper's torch and survive today. *Leon Callaway*

Tender trading on Union Pacific had been common for decades (as it was on many roads, large and small). It occurred many times across classes as capacity requirements changed over time, or for convenience if a suitable tank was available while its locomotive was undergoing heavy maintenance. In the case of the 4000s, company records suggest they stayed within the class, but it did not take long for late tanks to be found with early 4000s and vice versa.

Number 4014's tender, paired with the engine since the late 1950s, carries the number plate 25-C-116, indicating that it was the tender delivered with No. 4015. To make the switch, all that needed to be done at the shop was to change the last digit of the locomotive number painted on the rear of the tank.

Big Boy experiments

Overall, the design of the Big Boys proved successful, but that did not stop Union Pacific from conducting experiments with the locomotives. In the mid-1940s, a number of trials were begun but no permanent changes were made to the 4000s. In late 1945, the 4019 was fitted with smoke wind wings similar to those applied to Challengers that were assigned to passenger pools on the Los Angeles & Salt Lake and Oregon-Washington Railroad & Navigation lines. Always looking for improvements, Union Pacific tried exhaust stack experiments, with No. 4007 fitted with a single stack about 1947. Unfortunately, the records did not survive to tell the full story, but the 4000s retained their double stacks.

Perhaps the most controversial experiments were oil-firing trials conducted with No. 4005 in late 1946. The long-held presumption that the oil firing with No. 4005 was somehow a failure should be put aside. The dismaying aspect of a Big Boy using oil fuel was the huge quantities it consumed and the logistics of having heated bunker C oil accessible in all the places an engine might need it. Logistically, once the trials showed just how much oil was being consumed, it was far simpler to continue to burn company coal from Hanna, Wyo., or Rock Springs, Wyo. The situation with passenger power was far different. Engines in that service did not consume nearly as much oil and were able to easily run between terminals without refueling stops. Anecdotal evidence suggests that No. 4005 fired well on bunker C, using steam heat in the oil tank to keep its approximately 6,000 gallons of oil viscous. Unfortunately, as with the stack experiments, company records of those trials apparently disappeared more than five decades ago.

Throughout 1946, changes were applied to the 4000s' reverse gear and valve motion. Their 12-inch type H power reverse was replaced by the 24-inch type M. Additionally, needle bearings were applied to all jaws in the valve gear, and the eccentric cranks were fitted with roller bearings.

Postwar operations

Once World War II traffic began to decline, the 4000s spent more time on the eastern end of their territory. The postwar years saw Big Boys moving long troop trains across Wyoming. One occasion saw a 4000 with 49 sleeping cars headed east from Rawlins.

By 1948, Big Boys were making fewer trips to Ogden and were operating regularly east of Green River to Laramie and Cheyenne, and soon could be seen going south to Denver. As the mid-1950s approached, gas turbines and huge numbers of EMD GP9s were taking over in western Wyoming, and the 4000s kept to the east end of their range. Some 4000s were used with snow removal equipment after a 1949 blizzard, and although not common, by 1950 they were also seen working eastward to North Platte.

The Big Boys continued to put on an amazing performance, wrestling 4,450 tons on the ruling 1.14 percent (and about 5,900 tons with a Mikado helper). On the 1.55-percent Sherman Hill ruling grade through Buford, their rating solo was 3,250 tons. After the 1953 completion of the low-grade third main track on Sherman Hill, to the south of the original route, their westbound tonnage rating increased to 6,000 tons. In busy times, both routes were kept full with traffic, and it was common to see doubleheaded Big Boys westbound, sometimes just to transfer power west. But it was not to last.

The increasing number of diesels on the UP system throughout the 1950s meant less work for steam power. According to Lloyd E. Stagner's book, *American Steam Finale, 1954-1970,* steam freight ton-miles decreased from 56.5 percent in 1953 to 31.2 percent in the first quarter of 1954.

The arrival of 100 additional GP9s on Union Pacific in 1957 sidelined steam across the system, and No. 4005 made the last steam trip for the year across Sherman Hill on November 5. The next year saw 10 Big Boys in service between Cheyenne and Laramie, starting August 22. In 1959, UP decided to shop six Big Boys for service. They entered Cheyenne-Laramie service on July 6, but a steelworkers strike on July 15 and weak California fruit harvests meant UP steam's last gasp was strikingly brief.

At 10:15 p.m. on July 20, 1959, UP engineer Bruckert eased No. 4014 out of Laramie for its last revenue trip over Sherman Hill. It arrived in Cheyenne at 1:50 a.m. on July 21, 1959. A few hours later, No. 4015 made the final trip of a Big Boy, tying up in Cheyenne at

Where The Big Boys ran

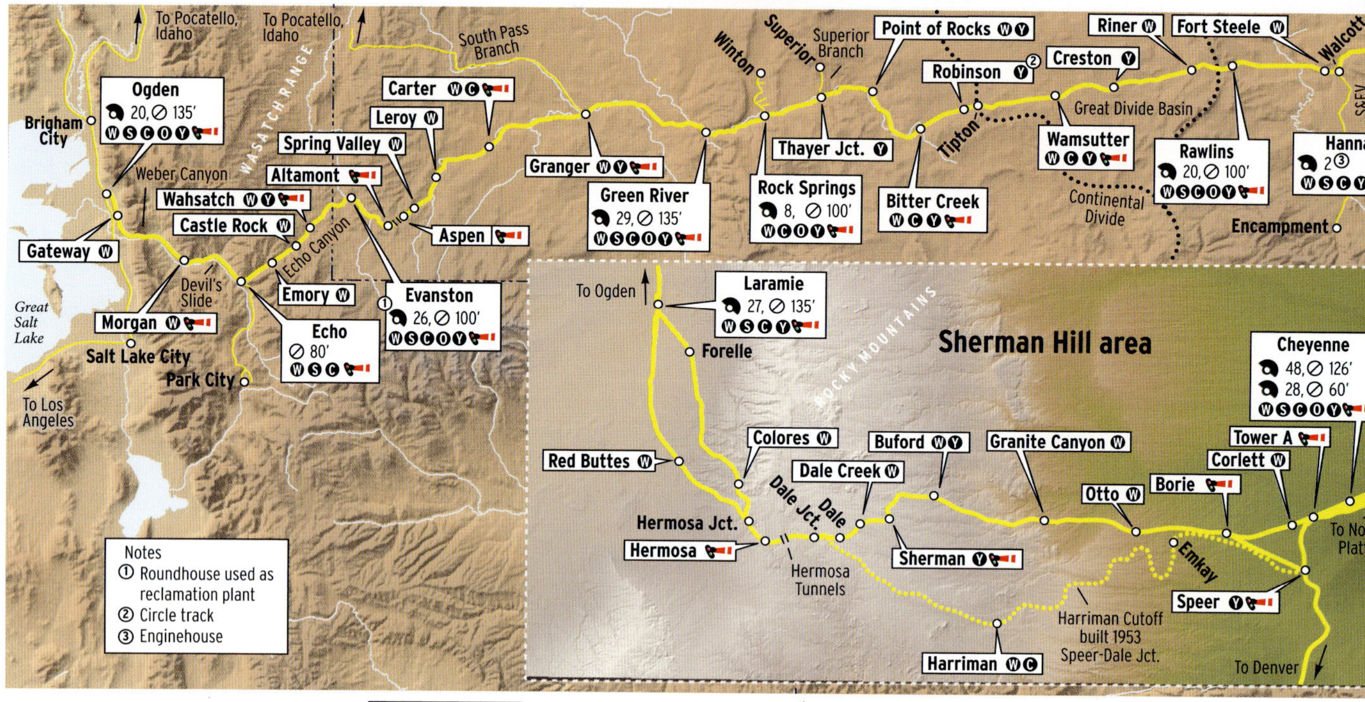

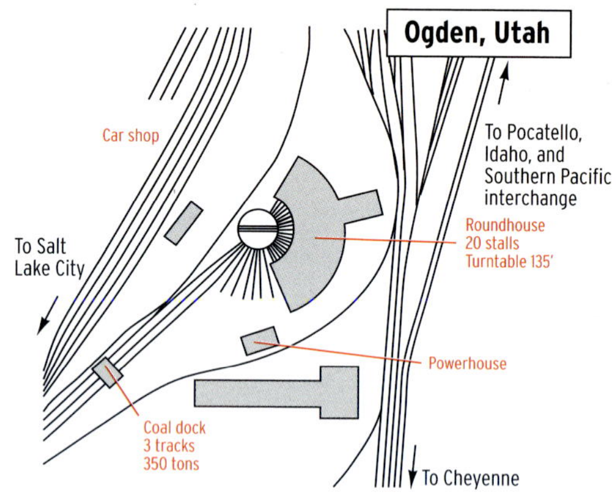

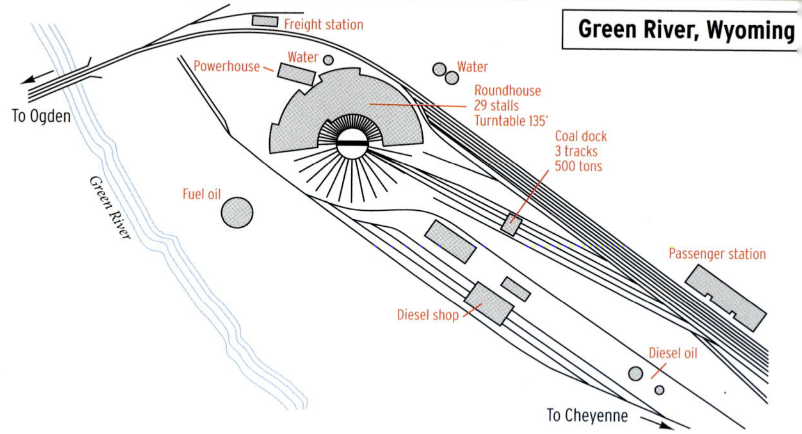

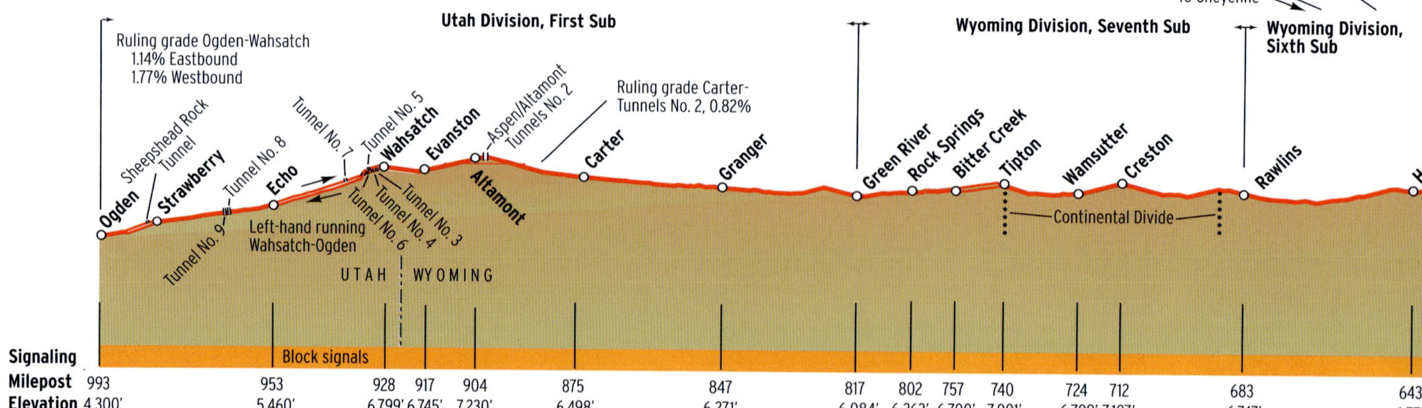

Miles are measured from Council Bluffs, Iowa; elevations are according to UP track charts, revised 1934.

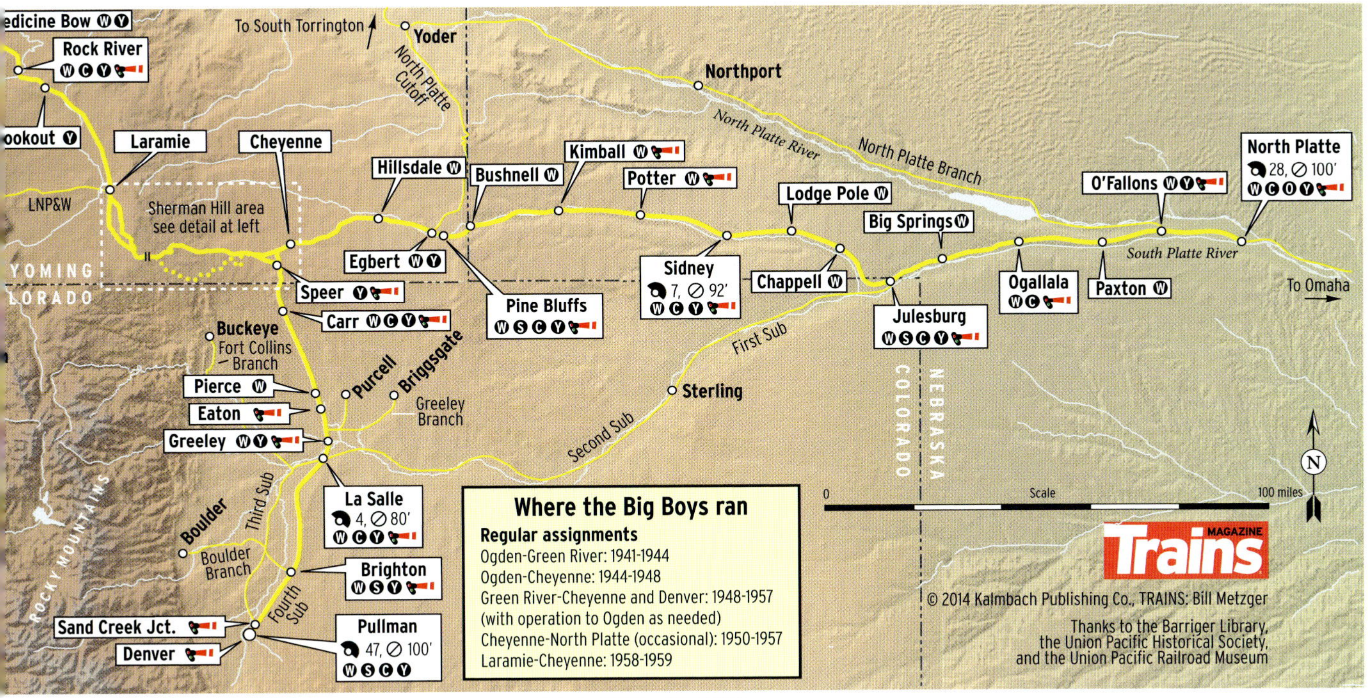

Medicine Bow W Y
Rock River W C Y
Lookout Y
Laramie
Cheyenne
To South Torrington
Yoder
North Platte Cutoff
Northport
North Platte River
North Platte Branch
North Platte 28, 100′ W C O Y
O'Fallons W Y
Hillsdale W
Bushnell W
Kimball W
Potter W
Lodge Pole W
Big Springs W
LNP&W
Sherman Hill area see detail at left
Egbert W Y
Speer Y
Carr W C Y
Pine Bluffs W S C Y
Sidney 7, 92′ W C Y
Chappell W
Julesburg W S C Y
Ogallala W C
Paxton W
South Platte River
To Omaha
WYOMING
COLORADO
NEBRASKA
Buckeye
Fort Collins Branch
Pierce W
Eaton
Greeley W Y
Purcell
Briggsgate
Greeley Branch
First Sub
Sterling
Second Sub
Third Sub
Boulder
Boulder Branch
La Salle 4, 80′ W C Y
Brighton W S Y
Fourth Sub
Sand Creek Jct.
Denver
Pullman 47, 100′ W S C Y
Rocky Mountains
Where the Big Boys ran
Regular assignments
Ogden-Green River: 1941-1944
Ogden-Cheyenne: 1944-1948
Green River-Cheyenne and Denver: 1948-1957
(with operation to Ogden as needed)
Cheyenne-North Platte (occasional): 1950-1957
Laramie-Cheyenne: 1958-1959
0 Scale 100 miles
N
Trains MAGAZINE
© 2014 Kalmbach Publishing Co., TRAINS: Bill Metzger
Thanks to the Barriger Library, the Union Pacific Historical Society, and the Union Pacific Railroad Museum

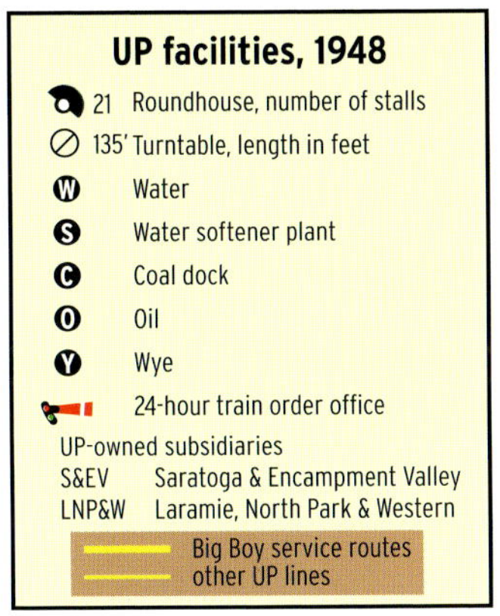

UP facilities, 1948
21 Roundhouse, number of stalls
135′ Turntable, length in feet
W Water
S Water softener plant
C Coal dock
O Oil
Y Wye
24-hour train order office
UP-owned subsidiaries
S&EV Saratoga & Encampment Valley
LNP&W Laramie, North Park & Western
Big Boy service routes
other UP lines

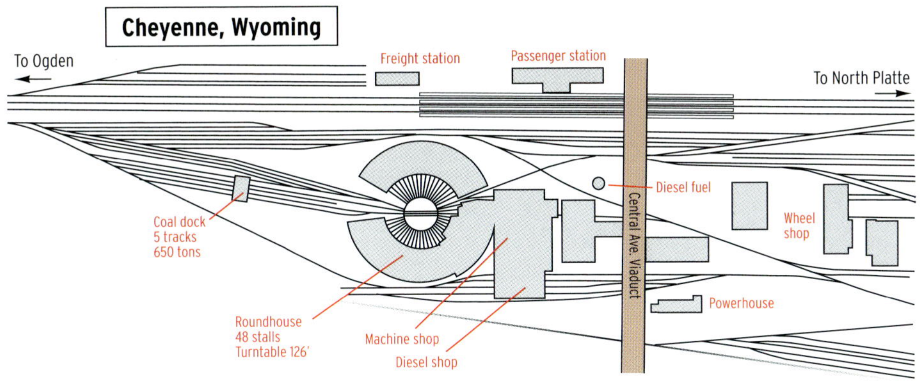

Cheyenne, Wyoming
To Ogden
Freight station
Passenger station
To North Platte
Diesel fuel
Coal dock 5 tracks 650 tons
Wheel shop
Central Ave. Viaduct
Powerhouse
Roundhouse 48 stalls Turntable 126′
Machine shop
Diesel shop

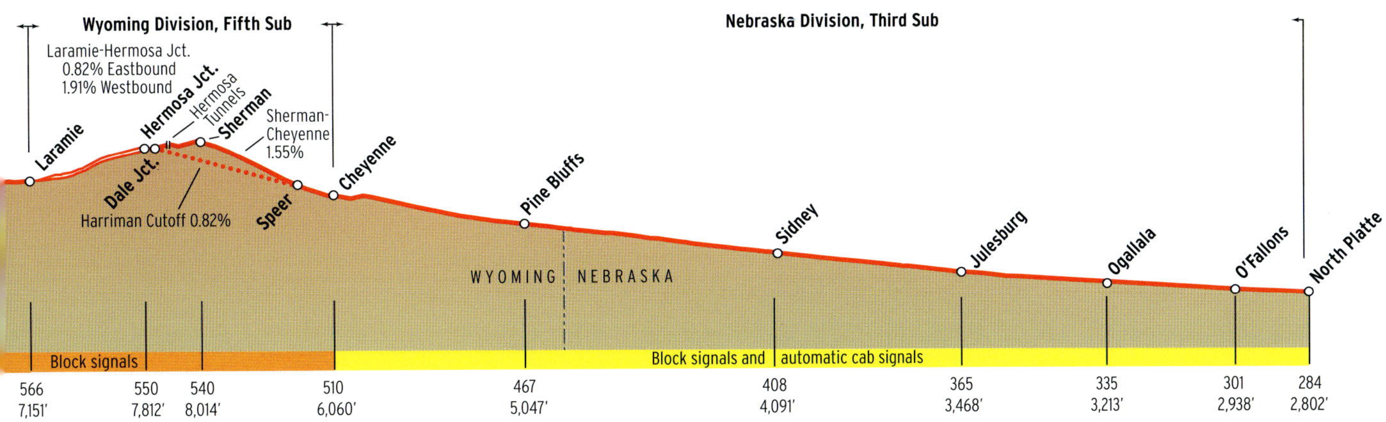

Wyoming Division, Fifth Sub
Laramie-Hermosa Jct.
0.82% Eastbound
1.91% Westbound
Nebraska Division, Third Sub
Laramie
Hermosa Jct.
Hermosa Tunnels
Sherman
Sherman-Cheyenne 1.55%
Dale Jct.
Harriman Cutoff 0.82%
Speer
Cheyenne
Pine Bluffs
Sidney
Julesburg
Ogallala
O'Fallons
North Platte
WYOMING NEBRASKA
Block signals
Block signals and automatic cab signals
566 7,151′
550 7,812′
540 8,014′
510 6,060′
467 5,047′
408 4,091′
365 3,468′
335 3,213′
301 2,938′
284 2,802′

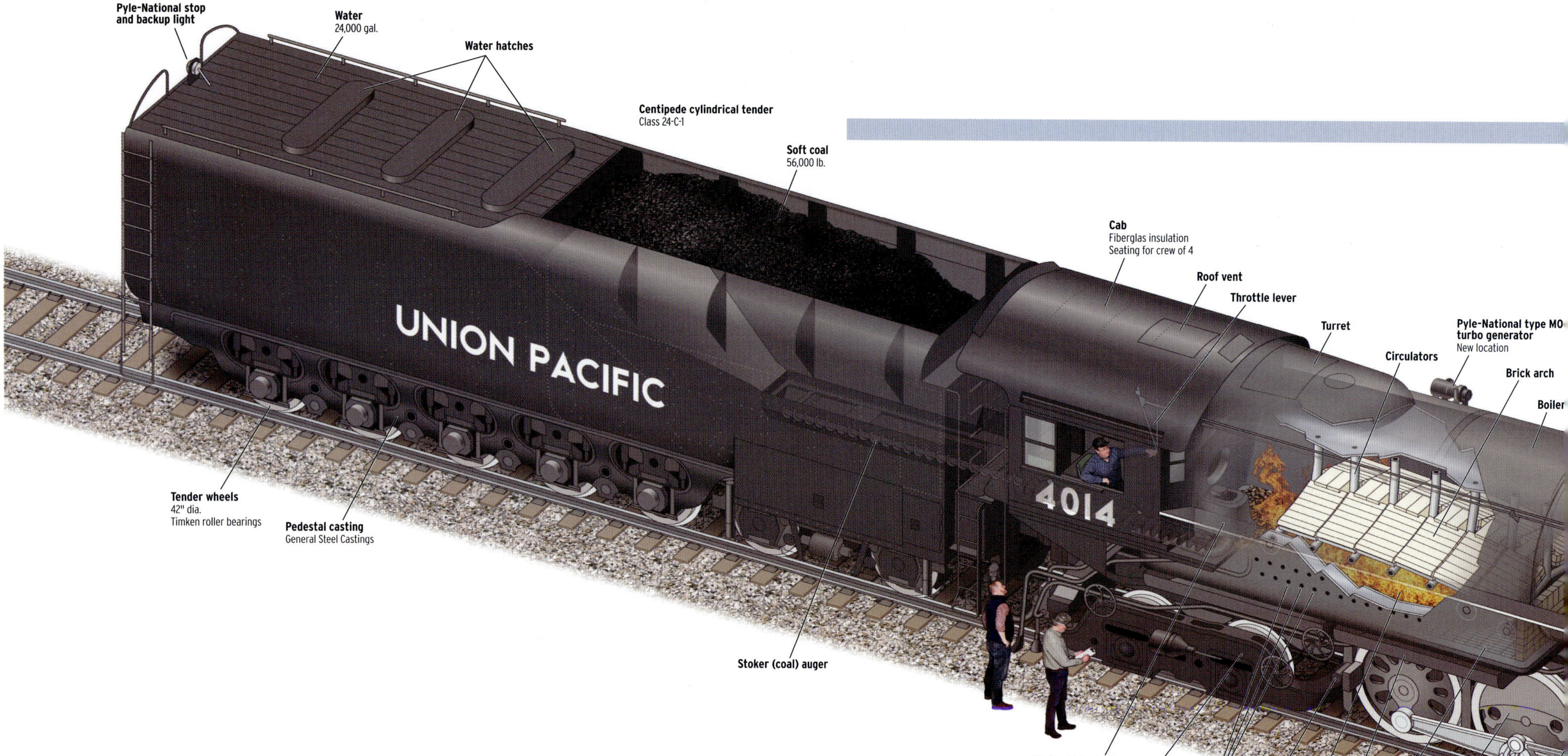

WHEN UNION PACIFIC train crews first looked at Big Boy No. 4014 in 1941, they had to be impressed. Before them was a 600-ton, 132-foot, 7,000-hp machine designed to climb long mountain grades with 3,600-ton freight trains and flat-out run at 70 mph. Inside and out, Big Boy is an over-the-top bit of rolling genius. With this cutaway drawing you can follow coal and water as they're converted to 300-psi steam that is used to move four pistons to drive 16 68-inch-diameter Boxpok-type drivers. Note the carefully constructed package, designed to fit into the railroad's profile. *— Jim Wrinn*

Big Boy: Inside and Out

Cutaway drawing shows the guts behind the king of steam's glory

Illustration by Rick Johnson

7:55 p.m. The arrival of Challenger No. 3713 on July 23 marked the end of regular service for UP steam.

The locomotives remained on the roster, yet languished in storage in Wyoming for another two years before they were officially retired and sent to scrap starting in 1961 and 1962. Fortunately, Union Pacific officials, employees, and railroad and locomotive enthusiasts prevailed to save some of the railroad's steam power for museums, in parks, and as community shrines.

Saving the Big Boys

Thanks to their popularity, eight of the 25 Big Boys were saved (see Chapter 10), landing in places ranging from the appropriate — a park a few blocks from the Cheyenne shop — to the almost inconceivable — a private steam museum in far-away Vermont.

About the same time, Union Pacific began running its last new steam passenger locomotive, 4-8-4 No. 844, on special excursions, thus keeping the company's rich heritage alive for future generations.

Why No. 4014 was saved, and not some other member of the class, is unknown. What we do know is that the Southern California Chapter of the Railway & Locomotive Historical Society was one of the organizations that asked to save a Big Boy. The club wrote to the Union Pacific to request the donation of a Big Boy, and the railroad agreed, preparing No. 4014 in Cheyenne and shipping it to California in late 1961 and early 1962 in a regular freight train.

A train order documenting the move was recorded to say:

"Santa Fe - JAN 6 1962 - TRAIN ORDER NO. 30 - To Conductors & Engineers EXTRA UP 165 WEST - At BARSTOW - SPEED LIMIT 20 MPH WHILE HANDLING UP STEAM ENG 4014 IN YOUR TRAIN - A K J [A.K. Johnston, Superintendent] - Complete 805 AM - Myers Opr."

Extra 165 West left Barstow, Calif., with six Geeps on the point. Number 4014 was 31 cars back, straddled by three gondolas ahead and three behind for braking. As was done when 4014 was delivered in 1941, all side rods were in place, but the pistons and piston rods were transported in the tender so as to prevent lubrication issues.

The freight cars ahead of No. 4014 and her braking cars were set out at UP's yard in Colton on the way to La Puente, Calif., where the locomotive was turned over to Southern Pacific for the short move to Bassett, Calif.

The 4014 heads west out of Cheyenne on Jan. 2, 1956. The locomotive would make its final revenue trip, from Laramie, Wyo., to Cheyenne, on July 20-21, 1959, just two days before Union Pacific ended all regular use of steam power. *James L. Ehernberger*

Urban legends of the Wasatch

The Big Boys were denizens of some of the most remote places in America, but they still generated their share of urban legends.

One such story involves Adolph Hitler, whose spies are said to have told him of huge locomotives, capable of handling heavy trains through the Rocky Mountains at high speed. It's undeniably intriguing to think that history's greatest villain might have obsessed over this symbol of American industrial might, but it doesn't seem to be true.

Steve Lee, former head of Union Pacific's steam program, heard stories about German prisoners of war passing through Cheyenne, allegedly awed and dispirited at the sight of doubleheaded 4000s charging forth to do battle with Sherman Hill. One of the captives was heard to mutter that, "any nation that could build something like that cannot be defeated," according to the story. But as Lee points out, the 4000s didn't operate out of Cheyenne during World War II; they ran on the Wasatch grade between Ogden, Utah, and Green River, Wyo. Nor could this be a case of mistaken geography: The Big Boys didn't run as doubleheaders between Ogden and Green River, owing to the tunnels.

Additional circumstantial evidence comes from the wartime files of the FBI. Eight Nazi saboteurs were sent to wreak havoc on America's war-making capability in 1942 (see *Classic Trains*, Winter 2001). The agents were captured before they could carry out their mission, but we know from their confessions that their rail objectives included Horseshoe Curve; Penn Station in Newark, N.J.; Hell Gate Bridge; unspecified targets on the Chesapeake & Ohio; and Great Northern's Cascade Tunnel. The sabotage operation was person-

ally conceived by Hitler, which suggests that the targets would have included Union Pacific if the Führer had been interested in the Big Boys.

Another persistent legend holds that No. 4005 was sold to Argentina. In one book, the locomotive is even said to have been "partially dismantled" in preparation for the move. The 4005 did earn distinction as the guinea pig for various modifications (up to now, it remains the only Big Boy converted to burn oil, for example), but there was never anything to the Argentina rumor. The 4005 has spent its retirement years at Denver's Forney Museum of Transportation, one of eight Big Boys donated to various parks and museums. The other 17 were destroyed — but intentionally so, not as the result of sabotage. — *Peter A. Hansen, author and editor of the Railway & Locomotive Historical Society's publication,* Railroad History.

Inset above: No. 4014 rolls under semaphores at Echo, Utah, with a merchandise train. Below: No. 4014 traverses the Harriman Cutoff, also known as Track 3, the preferred westbound line across Sherman Hill after 1953. *Above, Union Pacific; below, James L. Ehernberger*

On Monday morning, Jan. 8, 1962, Pacific Electric moved No. 4014 to the Los Angeles County Fairgrounds at Pomona, where it joined another steam locomotive unique to the Union Pacific, the only surviving 4-12-2, No. 9000, placed there in 1956.

Over the next few years, the surviving eight Big Boys stood proud but quiet in their geographically spread-out locations, their years of toil in the Utah mountains and Wyoming's high plains a fading memory to those who had maintained, operated, or were simply fascinated by them.

Number 4014 itself moved only once more, being pushed and pulled across the parking lot in 1989 when the RailGiants museum was relocated within the fairgrounds. It and other pieces of rolling stock in the collection were moved on panel track, and at one point, those in charge of the move accidentally let the engine roll free, and it pushed a Santa Fe Hudson off the end of the rails. Seeing this, and in retrieving the Hudson, some noted how freely the 4-8-8-4 rolled, and those who favored the 4014 said it was a sign that the old locomotive wanted to go once more.

But as the years went by, even the most optimistic of steam locomotive fans around the world agreed that a Big Boy would never run again. Restoring a Big Boy was the stuff of wild imaginations. It would cost a fortune to rebuild, take years to accomplish, and then, where could such a massive beast run? No, a UP 4000 had turned its last wheel under steam in 1959, they concluded, and that was the end of the Big Boy story. Or so they thought.

Union Pacific had other ideas.

In 2012, the railroad approached the Southern California group about reclaiming No. 4014. When the club agreed, UP painstakingly began inching the Big Boy out of its Pomona home in late 2013, and in April and May 2014, carefully moved the locomotive to Cheyenne for restoration. That process was completed just in time for the Golden Spike's 150th anniversary.

And so the legend, born in 1941, was reborn in 2019.

This article appeared in Big Boy: Back in Steam, *a* TRAINS *special edition, in 2019. Portions of this story were adapted from Gordon McCulloh's writings on the history of the Big Boy locomotives that appeared in the Fall 2013 issue of the Union Pacific Historical Society's publication,* The Streamliner.

It's late in the steam era, June 28, 1956, to be exact, and No. 4014 is doubleheading with three diesels (two Geeps sandwiching an F unit) eastbound at Dale, Wyo. A portion of the Dale Creek fill is visible to the left of the locomotive.
James L. Ehernberger

35

2 Big Boy Impressions

by David P. Morgan

Engine crews loved the 4-8-8-4s

BIG BOY — WAS EVER A RAILROAD NAME MORE APROPOS? An Alco workman chalked that on the 4000's bald smokebox door in 1941 when the first of Union Pacific's 4-8-8-4 single-expansion steam locomotives, still in the raw, naked, unpainted metal of her birth, moved out of the Schenectady, N.Y., plant with a fire on her grates and steam in her dome. Just "Big Boy" in chalk capital letters, plus a "V" for victory in the war that was only weeks away. But the name took hold, went into the railroad lexicon to stay for as long as men would talk about steam power.

And Big Boy she is on an October morning in 1957, in Cheyenne, Wyo., looming up on a yard lead for Extra 4008 West. She's immense, yet negotiable. She weighs in, engine and tender fully loaded, at 1,189,500 pounds — say, 594-plus tons. Seven times the weight of UP's original 4-4-0s, yet riding rails the same distance apart and capable, for all her brute strength, of moving faster. From those rails to the rim of her twin stacks it's 16 feet, 2½ inches; between the couplers protruding from her Commonwealth pilot and her 28-ton, 24,000-gallon-capacity tender it's exactly 132 feet, 9⅞ inches.

How big? Union Pacific says cautiously that the 4000s are "the largest in size and heaviest in point of total engine and tender weight of any simple articulated eight-coupled locomotives which have been built." In point of engine weight alone, Big Boy was eclipsed by the World War I Triplexes as well as by latter-day Chesapeake & Ohio and Norfolk & Western steam turbine-electrics (and almost shaded by C&O and Virginian's six-coupled 2-6-6-6s, oddly enough). Moreover, quite an assortment of reciprocating engines, all of them with smaller driving wheels, topped the 4000's tractive effort (see Chapters 4 and 5). But in overall size, lumping engine and tank, no other reciprocating steam engine equaled Big Boy. How big? Well, when you're sitting in a 4000's cab your feet are about 8½ feet above the rail.

ENGINEER JOHNNY FLOHR wraps a glove around the brass squeeze-type grip of the 4008's throttle at 11 a.m., and with the steam gauge needle quivering between 290 and 295 pounds, the extra pulls. Behind the tank are 103 empties (mostly Pacific Fruit Express reefers deadheading west) and three loads — just 3,000 tons, or half of what the 4-8-8-4 could haul over Sherman Hill.

Tower A gives the extra a green-over-red to enter the main, a leaky injector causes the lead engine to slip on the turnout (which closes the throttle, which opens the pops), then quickly, confidently 4008 is getting into stride, raining the countryside with black blots of cinders, talking it up with a deep-throated roar — really a bellow, but in rhythm nevertheless.

Big Boy swings onto Track 3 — the new line completed in 1953 that shaved Sherman from 1.55 to 0.82 percent — and settles down to 22- to 25-mile-an-hour operation. The throttle is wide open, the reverse notched up. Not notched up so high as to cut the fire, of course — which is to say the even, white-hot layer of burning coal over the grates. It's said that 70 percent of the black stuff spewed in by the stoker never hits the grates at all — it simply burns in air in a suspended state.

"Highball!" The blinking yellow of a searchlight signal ahead in a cut changes to green

Big Boy No. 4008 climbs Sherman Hill westbound in October 1957. *Philip R. Hastings*

Extra 4008 West, October, 1957

That's 85 feet, 9½ inches of engine reaching from the edge of the cab roof forward to the pilot. *Three photos, Philip R. Hastings*

The couplers match, but the tender of 4008 at left overshadows almost all other dimensions of the small tank car at right.

The 4008's Cheyenne departure shakes the ground, full of sound and fury, a black, belching, rolling monster.

Extra 4008 West, October, 1957

Walschaerts working: Rear engine of 4008 contributing half of tractive effort at 30 miles per hour. *Three photos, Philip R. Hastings*

Her job done, 4008 idles downgrade into Laramie past diminutive 2-8-0 No. 535.

Into Laramie steams Big Boy trailing 3,000 tons. Empty Pacific Fruit Express reefers trail into distance.

Unnumbered but not unsung, Big Boy 4000 is christened in chalk at Alco in 1941 with the name that stuck. Seventeen years later, 4000s were concluding their stint of fall 1958 service. *American Locomotive Co.*

and fireman Jake Kagel lifts a gauntleted hand in acknowledgment. Ahead of his green leather air-cushion seatbox, seemingly a mile of thin, cinder-spattered running board stretches taut along Big Boy's boiler (actually it's about 65 feet). Into the cut, now, with sunlight revealing the silhouette of twin stacks and smoke, steam swirling from pops and turret-valve housing. In lineside pastures horses stop, stare, then return to munching (and what does a horse think anyway of this black, belching, rolling monster that pounds so methodically by?).

Big Boy's cab is expansive — 10 feet, 10 inches wide over cab handholds, and large enough inside for four cab seats plus room for a card table with four places in the area between the raised ramps that accommodate the seatboxes.

The backhead is a mass of gauges, levers, valves, cocks — regulating or recording everything from the dynamo to the drivers. In practice, however, there are just a few essential items that insist upon constant check or manipulation. The top water gauge, for instance; the glass should show at least a half-inch of water while the engine is climbing a grade so that the front-end flues are covered with water and not scorched. Also, the exhaust steam injector encloses a glass opening marked Spill. When it illuminates, the meaning is that the cold water pump is functioning but not the hot water pump because an air bubble is fouling the vacuum in the line. Normally, an alert fireman could spot such a condition by peering down from the cab. But not in a storm; hence the light.

Big Boy keeps pulling, obeying the demands of the open throttle, the engineer's fingers playing on the sanders, the fireman's right hand on the injector — keeps gulping down coal and water and transforming both into ton-miles per freight-train hour.

HOW DID A LOCOMOTIVE like the 4008 come to be? What caused the construction of a "world's biggest?"

It all began in 1940 under the direction of Otto Jabelmann — the late, much-remembered Jabelmann, born in Cheyenne, raised and reared Union Pacific from crew caller to roundhouse foreman to vice president. It was Jabelmann who broke with the concept of the articulated as a low-speed compound to produce the high-speed single-expansion 3900s in 1936 — the first 4-6-6-4s ever built — also Jabelmann who ushered in the 800s, UP's fleet of 45 100-mile-per-hour 4-8-4s.

There was in all of this a grand design, a pattern, an outlook. Union Pacific thinking was geared more to horsepower than to initial tractive effort and tonnage ratings. From a drag era of railroading characterized by 2-10-2s and 2-8-8-0 compounds, with 63- and 57-inch drivers respectively, UP moved — as early as 1926 — to the longest rigid-frame engine ever built for U.S. service, a three-cylinder 4-12-2 with both a four-wheel engine truck and 67-inch drivers to accommodate speed. This locomotive, producing 80 percent more ton-miles per hour than a Mallet (and on half the coal, figured on a tonnage-moved basis), persuaded the road that big trains and speed were not necessarily incompatible. Moreover, Union Pacific — then as now — did have comparatively more perishable and manifest lading and easier grades than most Western carriers.

Other roads did not agree at the time and some still shudder. Southern Pacific, for one, never put a driver larger than 63½ inches under a purely freight locomotive.

The problem put to Jabelmann was this: Design the biggest, most powerful locomotive possible within existing clearances and axle loads of 67,500 pounds, design it to move 3,600-ton trains east out of the Great Salt Lake Valley from Ogden, Utah, to Green River, Wyo. (ruling grade: 1.14 percent) without a helper, and design it for safe operation at speeds up to 80 miles per hour.

THE ORIGINAL ENGINE of the specification, No. 4000, was in service little more than two years after the design study began and less than nine months after American Locomotive Company got a firm order for 20 of them. The problems conquered were in keeping with Big Boy's size. Weight, for example. Boiler barrel sheets of 1⅜-inch thickness were necessary, a figure for which there was no precedent. Despite such sizes, actual construction proved the weight to be within 1 percent of that calculated on paper, and yet on Big Boy 1 percent is 7,000 pounds — which had to be adjusted to control the weight on drivers to safe limits.

Articulating such enormous weights sprung fresh questions. How do you push Big Boy through a curve so that wheels do not tend to lift or bind, and how do you compensate for vertical movements encountered in change of grades without upsetting the weight distribution of the machine?

The solution was Alco's "lever control," defined by the railroad in these words:

Use of the term "lever principle" arises from employment of a definitely selected pivot point in the locomotive wheelbase about which the mass of the locomotive rotates with respect to track as the locomotive passes around curves. On an eight-coupled driving wheelbase this is the rear pair of driving wheels in which no provision is made for lateral movement of the axle in respect to the engine bed. The guiding wheels on the front engine unit (the front pair of truck wheels and the front pair of driving wheels) have provision for ample lateral movement against controlled resistance. The initial resistance of these wheels is about 17 percent, increasing gradually as the movement progresses. The second and third pair of driving wheels adjust themselves freely against a somewhat lower initial resistance and through a somewhat less range of lateral movement than that effective on the guiding wheels. Wheels back of the pivot pair control rear end movement of the locomotive against an initial lateral resistance somewhat lower than that of the guiding wheels. All wheels are fitted to track gauge with a setting of 53⅜ inches between backs of the tires.

The effect of this arrangement is to produce a rigidly guided locomotive when on tangent track that adjusts itself freely on curves with a guiding force cushioned in its application. Locomotives ... move around curves smoothly with complete absence of the succession of violent guiding oscillations characteristic of many existing steam locomotives.

In counterbalancing, advantage has been taken of absence of the tendency to nose brought about by lateral rigidity of the wheelbase on tangent track to keep down the overbalance, which has been fixed conservatively at 28½ percent of the reciprocating weights.

To relieve tendency of locomotives with long wheelbases to overload the driving springs when passing over concave vertical curves at summits and to underload them with corresponding overloading of the truck springs at the ends of the wheelbase when passing over a convex curve, an unusual degree of flexibility has been provided in the spring rigging by the employment of coil springs at all points of anchorage of spring rigging to the engine bed and trailer truck. Each of these cushion springs comprises two 8-inch double

The builder's photo of No. 4022, a member of the wartime series of five Big Boys delivered in November 1944, illustrates how the huge firebox was mounted on a trailer truck as well as two driving axles. *American Locomotive Co.*

coils in tandem; they permit the elongation or shortening of what are customarily hangers of fixed length, and permit vertical adjustments throughout the entire wheelbase to conform to track with a minimum of distortion of the adhesion weight on the driving wheels.

The locomotive as a whole has a three-point suspension. All driving wheels of the front engine are equalized on each side and the two sides are cross-equalized at the front and to the suspension of the rear end of the main equalizer beam, the front end of which bears in the Bissel type center pin of the engine truck. Each side of the rear engine is equalized as a unit from front to back, including both trailer wheels.

Of course the 4-8-8-4 wheel arrangement, a natural elongation of the Challenger, made for a natural balance in engine design. There was the hurdle inherent in any large

Gross ton-miles on Green River-Ogden line

YEAR	WESTBOUND	EASTBOUND
1939	1,273,156,000	1,494,615,000
1940	1,346,882,000	1,621,380,000
1941	1,848,204,000	1,987,939,000
1942	2,518,438,000	1,362,290,000
1943	3,357,425,000	2,601,525,000

A Big Boy boiler is swung over frames during construction at Alco's Schenectady Works in 1941. "Put a little warm water in 'em and they'll go" was the verdict of a man who fired them. *American Locomotive Co.*

Detail photos show the front, including the rarely used retractable coupler cover on the cast pilot (left), and the cross-compound air compressor tucked above the lead truck (right).
Four photos, Union Pacific

The Big Boys used Walschaerts valve gear and rode on Boxpok (pronounced "box-spoke") drivers (left). The rear of the centipede tender was plain but massive (right).

articulated — that of squeezing a decent size firebox over 68-inch drivers on an engine 133 inches in overall height. But the task was done with ample ashpan capacity and air volume in a firebox measuring over 19 feet in length and incorporating 150 square feet of grate area.

Otherwise, No. 4000 and her sisters were, in detail and essentials, modern power in the sense of boosted availability and reduced maintenance. Four 36-pint mechanical lubricators, for example, were installed to reach 123 oil outlets, exclusive of pumps and tender; roller bearings were fitted to all axles; cylinders, guide yokes, and supporting crossties were built integral with the cast-steel main frames. In short, Big Boy was designed to produce high horsepower around the clock, to boost mileage between class repairs, and to expedite turnarounds at engine terminals.

Number 4000 was deadheaded west from Schenectady on New York Central and Chicago & North Western in early September 1941, churning up printer's ink with each mile. Fascinated by the key word "biggest," the press, by one count, gave Big Boy 521 mentions in papers in 45 states. From the standpoint of publicity, Big Boy was a natural — a titleholder minus the drawback of an experimental. As a locomotive, the 4-8-8-4 was simply a logical extension of the 4-6-6-4, and Alco work order No. 1844 covered 20 locomotives at a cost of approximately $250,000 each to prove it.

Number 4000 was set up in Omaha Shops, then she steamed west under her own power with some 100 cars. Finally she was in home territory — in Wyoming and Utah where super 135-foot turntables had been installed at Laramie, Green River, and Ogden to accommodate her bulk.

Big Boy arrived on time and just in time. Gross ton-miles on the 176-mile Green River-Ogden district of the main line swelled, then exploded under the successive influences of national defense and war (see the table on page 45).

What did Big Boy prove? Plenty

- So free-steaming were the 4000s that the mechanical boys in Omaha happily concluded that "even wheat farmers could fire 'em" — which was a consideration in the manpower-short war.
- There was a noticeable increase in pulled drawbars and snapped knuckles, not so much because of the 4000s' higher tractive effort (135,375 pounds vs. 97,350 for a 4-6-6-4) as because of the engineer's failure to wait for the air brakes to release on the tail end of the longer trains Big Boy made possible.
- Despite her comparatively large 68-inch drivers, Big Boy displayed an ability to get down and lug at 4 to 5 miles per hour — speeds at which the Challengers proved slippery.
- Placed in Ogden-Green River service, where the line up through the Wasatch Range has a 1.14 percent eastbound ruling grade (including 18 miles of it without a break), Big Boy took 3,800 tons east without a helper, covering the 176 miles regularly in 7½ hours. By 1944, the 4000s were saving UP the equivalent of three trains each way per day (or more than 1,000 train-miles daily) in this assignment, plus 1,690 helper locomotive-miles per day. They had hiked production of gross ton-miles per freight train-hour 7 percent and released 4-6-6-4s and their 2-10-2 helpers for service elsewhere.

In 1943, Union Pacific borrowed Santa Fe dynamometer car No. 29 and got, for the first time, an accurate report card of Big Boy performance. Among other things, the test engineers discovered that a wide-open 4000 could consume 100,000 pounds of water and 22 tons of coal per hour; and on April 3, 1943, No. 4016 produced a reading of 6,290 drawbar horsepower at 41.4 miles per hour. On the best of tests the 4000s, designed to haul 3,600 tons over a 1.14 percent grade and eventually rated at 3,800 tons, frequently managed to keep 4,200 tons moving at 18 to 20 miles per hour.

The supreme endorsement came in 1944 when the original engines, Nos. 4000-4019, were supplemented by five more, Nos. 4020-4024, of virtually identical overall specification. The latecomers were slightly heavier in the wartime absence of some of the lightweight metals used on the original set.

FROM THEIR MAXIMUM EFFORT throughout World War II to the fall of 1958, when they last saw active duty (though the last revenue runs of a Big Boy occurred in July 1959), the 4000s enlivened the landscape and paid dividends on their investment until overcome by technological change.

They underwent few changes — an occasional modification of drafting or nozzles, an abortive attempt to single the twin stacks — but nothing large. As a rule, the 4005 was the guinea pig and she is still remembered, by deed and number, for the 1½ years she burned oil. Recalls an engine foreman: "Man, you needed to haul an extra tank of oil. She just gulped it down." And a mechanical department man in Omaha remembers that the change was hard on flues — that she "leaked like a rainstorm."

Otherwise, the 4000s just kept piling up mileage — usually 7,500 miles a month, making

There was no room in the inn for the tails of four Big Boys parked in the Cheyenne roundhouse on an October evening in 1957. Weighing almost 214 tons fully loaded, the 4-10-0 tank of a 4000 totes up to 25,000 gallons of water and 28 tons of coal. *Philip R. Hastings*

A dental appointment for the 4015 finds a Cheyenne roundhouse worker attending to smokebox work. It's 16 feet, 2½ inches from the rim of that stack housing to the rail below. Unlike the 4-6-6-4s, Big Boy's headlight and number plate are off the smokebox door.
Philip R. Hastings

more than 50,000 miles between class 4 or 5 repairs (machinery) and over 100,000 miles on class 3 jobs (flues). In engine terminals they could be turned in as little as 2½ hours, and on the road a modern facility could put aboard coal and water in just 8 minutes.

In later years, they stole away from their original Ogden-Green River assignment to work out of Cheyenne to Green River, Denver and, rarely, North Platte, Neb. Big Boys were seldom permitted east of Cheyenne in moderately graded territory because the road couldn't "find enough cars for them to pull." Since all 4000s have steam lines for train-heating purposes, occasional troop trains marched up Sherman Hill behind Big Boys.

SO MUCH FOR THE MECHANICAL. What of the men who operated them, who rode them, who listened and watched?

"After all's said and done, they're quite an engine." So spoke an old brakeman one evening in the dome of No. 102 when a 4000 loomed in the distance.

"Best mountain engine we ever had." The word of an engine foreman who added, "You think she looks big? They look twice as big turned upside down in the ditch!"

"They can say all they want to about the diesels, but when it comes right down to it these s.o.b.s are hard to beat on the hill — put a little warm water in 'em and they'll go." That came from a fireman who spent the best years of his life on the left-hand seats of 4000s.

Compared with its smaller articulated kin, Big Boy shaped up like this in enginemen's eyes: a 3900 was faster and rode better; a 4000 pulled more, paid better, steamed freer.

Two minor notes are perhaps worth inclusion. A 4000 was not considered a good rotary snowplow engine. Snow removal is slow, tedious, back-and-forth work, and the air-operated cylinder cocks of a 4-8-8-4 tended to freeze up and render the locomotive inoperable until some brave soul crawled in between the drivers with a torch to warm up the air lines. And, upon occasion a 4000's crosshead guides have cracked, snapping the piston, dropping the main rod, and stripping the side of the engine. Today, there's a tiny weld joint on the guide. It doesn't support anything, really, but if an inspector sees it's cracked, the ominous meaning is that the tapered screws that hold the guides are working loose.

Oh, yes — one more item. That fancy retractable coupler — dreamed up so that a pilot presents a smooth appearance and doesn't grab engaged articles such as automobiles — doesn't. In actual practice it required three men, much sweat, and colorful talk to pull out a coupler, and the inevitable 20-minute delay just wasn't worth it.

BIG BOY HAS ALWAYS BEEN much more than just another locomotive, of course. The 4000s are a legend, perhaps the most famous engines since New York Central's 4-6-4 of 1927, certainly the biggest box-office star of any of the 3,000 articulateds that have operated in the U.S. Journalist Clyde Carley has ridden Big Boy for *True* magazine; 4000s have starred in color in *Time* magazine and in the late *Collier's* (which cited President Stoddard of UP as "Boss of the Big Boys"); and the 4-8-8-4s have occupied the cameras of TV's "Industry on Parade" and "Wide Wide World," not to mention an hour-long drama on "Robert Montgomery Presents."

In astute contrast to those dieselizing railroads who attempted to censor steam in its twilight years, Union Pacific has remained delighted with the reception granted its monsters.

It's nothing faked; it's something natural, obvious, inevitable, infectious. The p.r. men, the roundhouse workers, the white-collar clerks, the engine crews, virtually everyone on UP's payroll is Big Boy conscious, Big Boy proud.

Even today, in 1958, usually unemotional mechanical department men find it hard to completely write off the 4000s. As of June 1958, 17 were ready to roll in case traffic climbed and the long-delayed newcomers, the 8,500-horsepower gas turbines, remained at General Electric. The balance of eight "unserviceable" 4-8-8-4s were off the list for nothing more serious than tire jobs. A few were fired up in the fall.

Even if Big Boys cannot last forever, one seems assured of preservation. No. 4000 herself has long since earned the honor in publicity alone, to say nothing of ton-miles.

AT 12:33 P.M. EXTRA 4008 west halts beneath the coaling stage at Harriman, Wyo. Four minutes later, tender replenished, the 4008 eases back, collects 4 or 5 feet of slack, then begins to pull. Once more the leaky injector has sprayed the rails and undermined Big Boy's footing.

A thunderous roar ensues from her twin stacks, and 16 drivers plus 20 tons' worth of articulated rods give the giant locomotive a violent shaking. (Shutting off steam to stop slippage on a Big Boy takes guts; the engineer stands, braces himself, grips the throttle with both hands, and shoves forward mightily.) Once again the 4-8-8-4 is in stride, and as if asking forgiveness for her misdeeds, goes hammering upgrade with increased vigor and poise.

An example of how utterly dependent even a 604½-ton engine and tender is upon its

The coal pile in 4008's tender is dwindling as the 4-8-8-4 shouts through a 110-foot-deep cut on the new line over Sherman in August 1955 with an extra west of deadheading reefers.
Jim Shaughnessy

smallest component is served up minutes later when the stoker promptly quits cold. A locomotive capable of burning up to 11 tons of coal an hour fades fast without it, so a sort of controlled commotion, or educated panic, transpires in the cab. First, Engineer Flohr shuts down, hoping the trouble will be corrected before the momentum runs out. Next, the fireman lifts the stoker screw cover and reverses the mechanism, hoping that the foreign object will come to light. Finally — it seems hours instead of the seconds it takes — a 12-inch piece of knuckle pin inexplicably rides right through the stoker screw tunnel and into the firebox! Immediately Flohr widens out the throttle and 4008 has occasion to demonstrate her ability to lug.

All glory breaks loose as the 4-8-8-4 charges for the summit. Into one cut Big Boy blasts with the stacks belting out an indistinguishable beat akin to the pounding of a thousand rivet hammers, the sound being that of a super boiler shop in hi-fi.

The noise, the pounding, the deafening fight of the big engine on 0.82 percent with tonnage is so intense, so close, as to dominate the mind and heart at the moment of its happening. Down in this cut one suddenly has no family, taxes, appointments, U.N., deadlines, cares, joys — no nothing but the profound realization of being aboard the biggest thing in steam, being at the mouth of a boiler capable of delivering almost 100,000 pounds of superheated steam an hour to the engines below, being there when drawbar pull turns active instead of academic. All men are vain who have experienced this ... or should be.

Into another cut — and on this 42½-mile, $16 million new line they're deep. This one, near Dale, is 110 feet deep and the rock walls seem to preclude the exhaust from rising, seem to shove it back down on the ears, to the underplay of yoked drivers.

Green over red. At 1:07 p.m. the war is over as the 4008 comes stalking around a curve and through a crossover onto the original main line. It's Dale Junction.

At 1:11 the fireman slams cab windows and doors shut as the accelerating 4008 plunges into Hermosa Tunnel. The glass steams, the heat suddenly intensifies, cab gauges dim. Then into sunlight again and running free.

"Got her by the tail now," shouts Jake Kagel.

Stoker valve needles die off to zero.

Pressure declines to 275 pounds.

Throttle is closed.

The exhaust is gone (and what seemed so deafening is now suddenly missed, like an old friend) and the only sounds that remain are the sounds of rods and wheels, the

Extra 4019 West thunders into the dawn of August 24, 1958, near Borie, Wyo., as Big Boys talk it up in the twilight of their careers.
Richard F. Lind

The wonder of it all: Engineer Johnny Flohr looks over the mass of machinery he controls before taking No. 4008 west. *Philip R. Hastings*

occasional hoot of the whistle, the automatic air being cycled on and off — the monotone rumble accented by the metallic jiggle of cab pipes and gauges. Extra 4008 West is coming home as fast as Wyoming Division Special Rules No. 14 will allow it to roll.

A blue brakeshoe smog gathers about the trucks of the empty reefers behind as the train rambles downgrade to Laramie at 50 miles per hour.

Suddenly, on a blind curve, the fireman spots a herd of sheep, yells to the engineer, and pulls down a pair of fistfuls of air to simulate reaching for the whistle cord. Just behind 4008's stack housing the whistle erupts in short, urgent blasts. In a flash the herd separates at trackside, but for three sheep life is over.

The fireman scowls. "I don't like to do that."

Extra 4008 West catches a yellow eye on the outskirts of Laramie — actually a red over yellow. Brakeshoes take the roll out of 103 cars and 3,000 tons. The long engine eases on down, walks past the Union Pacific tie-treating plant, stops dead at 1:37 while the brake-man unloads to ring the yardmaster for a track assignment.

It's Track 7, and the 4008, bell ringing, train gently nudging her, inches on in. At 1:52 p.m. Extra 4008 West is dead in the yard at Laramie and, for all train dispatching purposes on the Wyoming Division, has ceased to exist.

Big Boy is cut off and taken to the roundhouse.

What was that the fireman said about Big Boy being hard to beat on the hill — "Put a little warm water in 'em and they'll go"?

Amen, brother. Amen.

DAVID P. MORGAN was the editor of TRAINS magazine from 1953 to 1987. This article first appeared in the November 1958 issue.

FAST FREIGHT hauler, No. 4015 races across Wyoming's high desert west of Laramie. This is Big Boy at its finest. *Robert Hale*

3 Without Equal

by Steve Lee

Union Pacific needed a bigger locomotive;
Alco provided the Big Boy

TO CALL A UNION PACIFIC 4-8-8-4 a "big locomotive" would be akin to calling the Mona Lisa a "nice painting." Built in two lots in 1941 and 1944, the 25 Big Boys were the acme of steam locomotive development for the American Locomotive Company, and were so successful they still made money for UP more than a decade after its managers had chosen to dieselize.

The Big Boy was the result of a design effort which had already produced two successful series of 4-8-4 Northerns and 4-6-6-4 Challengers. For the new 4-8-8-4s, the Alco-UP engineering and design team didn't need to break too much new ground — they only had to refine and enlarge what they'd developed and proven in service.

The design team started with a performance-based requirement for a locomotive that could pull 3,600 tons up the 1.14 percent ruling grade between Ogden, Utah, and Evanston, Wyo., without a helper — and worked up the design to meet those needs. The locomotive they produced exceeded expectations, as evidenced by successive increases in the Big Boy's tonnage ratings once experience revealed their capabilities. Ultimately, the Big Boys were rated at 4,450 tons on this district, 23.6 percent more than planned.

The Big Boys were big locomotives, so large that before the first one was delivered, UP had to invest money to remodel facilities at engine terminals where the Big Boys would be fueled, watered, and maintained. Turntables 135 feet long were installed at Ogden and Green River and Laramie, Wyo. Cheyenne, Wyo., got a 126-foot table. Several roundhouse stalls at each terminal were lengthened so the new locomotives would fit inside. Locations outside their usual territory, such as Denver and North Platte, Neb., were not equipped with the long turntables, and the railroad had to turn them on wyes, and use only roundhouse stalls aligned straight across the turntable from inbound tracks. At those locations, most 4-8-8-4 inspections and repairs had to be done outdoors. Water standpipes not already raised to clear the tall tenders on the 4-8-4s had to be raised so they could swing over the tops of the 4-8-8-4s' tenders.

Union Pacific's right-of-way also required investment to accommodate the Big Boys. The distance between adjacent tracks had to be increased on sharp curves so the smokebox overhang to the outside of curves would not result in sideswipes. Cuts on curves had to be widened for the same reason, as did curved tunnels such as between Devil's Slide and Morgan, Utah. Certain bridges and culverts had to be strengthened. Some trackage in terminals and yards had to be realigned. Even with all that, employee timetables carried long lists of speed restrictions specific to the Big Boys and they were kept from some tracks owing to their size and weight.

An Alco success story

Conventional wisdom holds that of the three steam locomotive builders, Alco, Baldwin, and Lima, the latter was the innovator, and Alco and Baldwin followed its lead. The same wisdom holds that Baldwin was the champion at building big locomotives in volume.

Where Alco differs from its competitors was its ability to build big locomotives in large quantities at a competitive price, and innovate. Of the three builders, Alco was the biggest proponent of three-cylinder power, and developed the ultra-high-pressure Delaware &

Union Pacific Extra 4015 West is in the yard at Laramie, Wyo., on Sept. 18, 1956. Laramie was a crew-change point in this era, and boasted a 27-stall roundhouse as well as a giant icing facility for refrigerator cars. *James L. Ehernberger*

Fireman Jake Kagel works the controls inside the roomy cab of 4-8-8-4 No. 4008 in 1958.
Philip R. Hastings

Hudson compound 2-8-0s and the triple-compound 4-8-0s. Lima's engineering staff and plant craftsmen certainly produced many successful designs; however, Union Pacific's Challengers and Big Boys demonstrate Lima had no monopoly on original thought.

While a few dogs emerged from Alco's factories, they were greatly outnumbered by success stories. Prior to, and during its collaboration with UP, Alco's design team collaborated with New York Central's engineering team to produce excellent 4-6-4 Hudsons, 4-8-2 Mohawks, and ground-breaking 4-8-4 Niagaras.

At the time, every huge railroad had its own engineering staff, some with almost as many mechanical engineers and draftsmen as the builders. While this produced a lot of waste, duplication, and "not-invented-here" prejudice, when an innovative builder was paired with a progressive railroad staff, successful locomotives resulted.

The Big Boy (like UP's 4-8-4s and 4-6-6-4s that came before) was a joint design effort of Alco's engineering staff and UP's Bureau of Research and Mechanical Standards. There's a surprising degree of standardization and parts interchangeability among the three types. Hundreds of drawings are common to all, with only a specific dimension here and there separating them.

While such standardization is to be expected with vendor-supplied appliances, here it extends to such items as basic firebox design; design of boiler courses; and seams, piping, and many running gear parts. The spring rigging, for example, designed (and patented) by Alco's Jerry Blunt, is almost identical among the three, with certain springs, pins, bushings, and hangers being fully interchangeable.

What worked, what didn't

UP's 4-6-6-4s of 1936 had upset some long-held theories. As did the 4-12-2s before them, the Challengers employed flat-bottom fireboxes and mud rings to keep overall size within turntable length limits without sacrificing grate area. This design also shifted weight to the drivers from the trailing truck. Critics have argued this firebox arrangement was inferior to the traditional slope design, but in practice, there was no difference in performance or costs between the two.

Too, the early 4-6-6-4s proved that articulateds could be more than a plodding drag freight engine. The Challengers ran efficiently at high speeds, utilizing their high-horsepower production capabilities without damaging the track or shaking themselves to pieces, yet still had high starting tractive effort. The excellent ride and tracking qualities of the Challengers (and Big Boys) was largely due to the four-wheel engine truck, which gave them great stability at speed and eased them through curves, switches, and crossovers.

Some decisions did not turn out as hoped. The first series of 4-8-8-4s, like UP's second series of Northerns, was equipped with Elesco Type E superheaters. This provided an increase in total heating and superheating surfaces as compared with the Type A, and promised reduced coal consumption. In practice, the Type E proved more costly to buy and had higher maintenance costs and failure rates.

An early problem with the Type E was burning of the return bends, which was eventually solved by shortening the length of the superheater units to get the return bends farther away from the firebox. Eventually, the units were shortened 23 inches, with a corresponding reduction in total superheating surface area.

Recurring headaches with Type Es convinced UP that the savings in coal created by the Type Es were more than offset by higher maintenance costs and down times. The second

Big Boy No. 4017 is on the turntable at Cheyenne in September 1954. Turntables, water standpipes, and roundhouses that Big Boys used along the locomotives' routings required modifications so the 4-8-8-4s could use them.
Jim Ehernberger; Al Chione collection

In this undated photograph, the front set of driving gear on a Big Boy 4-8-8-4 is hoisted aloft inside a steam shop enabling work under the frame. *Robert Hale*

series of 4-8-8-4s, the fifth series of 4-6-6-4s, and third series of 4-8-4s came with Type A superheaters and were better performers.

Another weakness was the exhaust steam injector. This infernal device, sometimes called a "poor man's feedwater heater" due to its lower cost, was used instead of the tried-and-true Worthington Type S or SA feedwater heater after experience with the first two series of Challengers, some of which had Worthington 6SAs. While the 6SA was probably the best one ever used by UP, its value was diminished by its need for a pump that was heavy and required mounting directly to a locomotive's bed frame. With the large diameter of the boiler on the Challengers and Big Boys, this was not feasible without major changes in the design of the locomotive, the pump, or both.

On the Challengers, the 6SA hot water pump was mounted in the only place it would fit, at the bottom front of the smokebox. This location made it difficult to access the pump for routine service and repairs, which were frequent due to unanticipated vibration. The vibration caused leaks and other problems including high stresses on the smokebox itself, both from the weight of the pump and from the action of its constant pumping.

A few other things were corrected. The exposed air pump aftercoolers on the front handrails didn't stand up to the vibrations inherent in their location and method of

Union Pacific 4-8-8-4 No. 4007 steams westward near Emkay, Wyo., in August 1958 with a string of reefers in tow. The time remaining in operation for these 25 iron giants could be counted in months. Eight avoided the scrapper's torch; No. 4007 did not.
Jim Ehernberger; Al Chione collection

mounting, and were vulnerable to damage during smokebox inspections and front-end repairs. They were replaced with Wilson aftercoolers located behind the air-pump shield out of harm's way, and affixed to better resist vibration.

The drifting throttle, used on descending grades, proved to be a primary cause of midnight creepers — unattended engines moving at engine terminals. These were removed. The actuating rods for the Nathan mechanical lubricators proved to be fragile, so they were replaced with a more sturdy chain-and-sprocket drive system that looked weird but was reliable.

Several changes were made in an effort to lower operating costs. Ten-inch steel sideboards were added to the tops of the tender coal bunkers, increasing coal capacity by 4 tons to 32 tons level full. The No. 4005 was converted to oil fuel in December 1946. It steamed well, but could not carry enough fuel to consistently make it from one oil tank to the next, because oil tanks were fewer and farther between than coal chutes. It was converted back to coal in March 1948. The No. 4019 was equipped with "elephant-ear" smoke deflectors for a short time in late 1945 and early 1946.

Nothing on earth like them

To the men who fired them, ran them, and maintained them, the Big Boys were simply the biggest and the best. Conversations with those who spent the early part of their railroad careers on and around the 4000s always elicit a tone of respect and reverence. But they seldom call them Big Boys; they were just "4000s."

If you ask them, these veterans will tell you how powerful the 4000s were. They'll tell you how in every dimension they seemed almost overwhelmingly large. They'll tell you how there was nothing louder, and how hot they were in tunnels — so much so that after several engine crews had the skin burned off their ears in hellish trips through the bores, the company issued leather hoods that covered the head. These were connected to a hose providing cool breathing air.

They tell of how the 4000s would steam even with entire grate sections burned out, broken, or missing. A small boy, said one veteran, could keep a Big Boy at 300 pounds of pressure, as long as he could reach high enough to operate the stoker valves.

They tell of a boiler so long that two water glasses, one above the other on each side of the cab, were necessary to keep track of the water level going up and down hills. They tell of cabs big enough to hold a union meeting, and a firebox so large it could, and often did, burn 22,000 pounds of coal an hour. Could you hand-fire if the stoker quit? No way, unless you were drifting downhill.

Did I mention they were loud? One retired engineer said that nothing on earth was louder than a 4000 working through a tunnel, to which another retorted, "Oh, yes there was! Two of 'em double-headed, especially if you were on the second one."

The pinnacle, then the fall

By the late 1930s, Alco was at the top of its game in the steam locomotive business. By 1941, American railroads, except for a few holdouts, had lost interest in steam. Orders for new steam locomotives dictated by wartime restrictions and increased traffic levels indicated to Alco managers a continuing market. It was wishful thinking — by 1944 even the UP wanted diesels, instead of the 35 4-8-4s, 4-6-6-4s, and 4-8-8-4s it received that year from Alco. UP would order no more.

Alfred W. Bruce, in his 1952 book, *The Steam Locomotive in America,* said the Alco 4-8-8-4 "... probably represented the maximum development of the articulated steam locomotive with a reasonable axle loading for manifest freight train operation, since it incorporated both four-wheel leading and trailing trucks to provide maximum boiler capacity and riding stability at high operating speeds."

STEVE LEE is a retired manager of train operating practices for Union Pacific. This article first appeared in the September 2001 TRAINS.

4 Compound Articulated, Simple Articulated

by Ed King Jr.

How developments in articulated locomotives led to the Big Boy

AS ONE ABSORBS THE WRITINGS about the history of the steam locomotive in America, a couple of things stand out. First, no commercial locomotive builder ever sold a locomotive that wasn't the most efficient and profitable engine that could be had, given the buyer's limitations of loading gauges and budgets. Second, nobody charged with obtaining power for a railroad was ever motivated by thoughts of buying a locomotive that wasn't absolutely the greatest thing that could be had. In other words, it's almost impossible to find a lemon — no one ever designed a lemon, no one ever built a lemon, and no one ever bought a lemon. Well, almost no one.

There are, however, some lemons that stand out in the literature if one looks hard, and both belonged to the Norfolk & Western. In 1910-1911, it bought or built 61 heavy 4-8-0 (Class M2) locomotives when just about everyone else was obtaining Mikados. The 2-8-2s were forward-looking designs and the 4-8-0 was a dead end; there wasn't room for an adequate firebox over those rear drivers, and the Mikado with its trailer truck had sufficient room. An N&W official would later say that the M2-class engines were too light for heavy work and too heavy for light work.

Another lemon was the Class K3 4-8-2, built in 1926 as an answer to Lima Locomotive Works' pioneering A-1 2-8-4 of 1925. Like the A-1, the K3 used 63-inch driving wheels. However, because of a long, heavy main rod connected to the third driver, the K3s were hopelessly under-counterbalanced, leaving the engines unable to attain the speeds suggested by their excellent boilers and fireboxes. N&W got rid of all 10 of the K3s during World War II (see "N&W's Nomad Mountains" in the February 1979 Trains for details).

From Alco's production in 1904 of America's first Mallet, Baltimore & Ohio No. 2400, a monstrous 0-6-6-0 for its time, up to the advent of Norfolk & Western 2-8-8-2 No. 2200 in 1952, authors have dictated the accepted and acceptable thought on the subject. The Mallet was a compound, expanding steam from the throttle into the two high-pressure cylinders to the rear; the steam was then exhausted into a receiver where it was channeled into the two low-pressure cylinders at the front. The steam expanded again in the low-pressure cylinders, and it was then exhausted through the nozzle and smokestack like a conventional locomotive. The piston area of the two low-pressure cylinders was larger than the high-pressure cylinders by a ratio of about 2.5-to-1 in order to compensate for the steam having been expanded once.

Compounds and clearances

Norfolk & Western had a problem: clearances. It obtained two classes of Mallets in 1910: five Class X1 0-8-8-0s from Alco and five Y1 2-8-8-2s from Baldwin. Both classes had 24.5-inch-diameter high-pressure cylinders and 39-inch-diameter low-pressure cylinders — the latter a limiting factor for all N&W steam. The X1s and Y1s used a 30-inch piston stroke and the 200-pound boiler pressure common for the day, and both classes rode on 56-inch driving wheels.

Such clearance problems didn't restrict cylinder size on other roads; N&W's neighbor Virginian Railway obtained some Class AD 2-8-8-2s in 1912 that had 28-inch high-pressure cylinders and 44-inch low-pressure ones. The Virginian would go on to obtain 10

On the Chesapeake & Ohio near Montgomery, W.Va., 2-6-6-6 No. 1647 blasts down the Kanawha River valley with 90 loads of coal in tow. The H-8 locomotives, the most powerful steamers ever built, came at significantly higher cost than similar engines on rival Norfolk & Western. *Phillip R. Hastings*

2400
2400

America's first Mallet, Baltimore & Ohio 0-6-6-0 No. 2400, rolls past the reviewing stand at the Fair of the Iron Horse near Baltimore in 1927 to celebrate the railroad's centennial. The locomotive, built in 1904, saw years of service on Pennsylvania's Sand Patch grade. *Baltimore & Ohio*

enormous 2-10-10-2s in 1918 with low-pressure cylinders of 48-inch diameter — the largest cylinders ever applied to a steam locomotive.

N&W's first Mallet fleet, begun in 1912, consisted of 190 fine 2-6-6-2s Alco built in 1910 with a design that incorporated superheaters and mechanical stokers. These Class Z engines got the N&W through World War I, but the railroad knew it was going to need something bigger for the future.

In 1918, N&W's Roanoke, Va., shop produced the Y2-class 2-8-8-2, which used the same cylinder dimensions and 56-inch drivers of the Y1 engines, but incorporated a longer piston stroke of 32 inches and a boiler pressure of 230 pounds. The Y2's power was thus equivalent to the Virginian's AD-class engines, and the Y2s were faster. The Y2 class eventually numbered 31 locomotives.

The United States Railroad Administration (USRA) was created in December 1917, taking over the nation's railroads to help them cope with the traffic glut caused by shipments of war materiel to eastern ports for shipment overseas during World War I. As part of its role, the USRA designed a series of standard locomotives to replace war-worn power on the nation's railroads. Twelve standard locomotive designs were developed by the USRA design team, and all three of the big commercial builders (Alco, Baldwin, and Lima) began manufacturing them in 1918. Locomotives were assigned to railroads based on need; many railroads continued buying and building USRA designs after the war.

One of N&W's mechanical engineers joined the USRA design team and obviously had a hand in USRA's standard 2-8-8-2. The USRA Mallet used a boiler pressure of 240 pounds and high-pressure cylinders of 25-inch diameter to go with low-pressure cylinders of 39 inches. It's interesting that the USRA Mallet was designed to fit N&W's clearances.

The N&W obtained 50 of the USRA Mallets, designating them as Class Y3, and dynamometer car tests found them to be superior to the Y2 engines. In response, the Y2s were modified with 25-inch high-pressure cylinders and had their boiler pressure raised to 240 pounds, after which they were equal to the Y3 USRA engines. The altered Y2s became N&W's Y2a class.

In the late 1910s, N&W had devised a large exhaust nozzle to be used in connection with a large-diameter smokestack in order to move the most exhaust gases with the least back pressure on the cylinders. Called the waffle-iron nozzle, it incorporated a central outlet and six annular ports to spread the exhaust to fill, or seal, the larger stack. This exhaust system would become standard on N&W's locomotives.

Not so simple to simple

The simple articulated — a locomotive built on the Mallet pattern but using live steam from the throttle in all four cylinders of equal size — first hit the rails in 1911 when the Pennsylvania Railroad built its sole HH1s-class 2-8-8-2. (In PRR nomenclature, the lower-case "s" signifies a superheater-equipped locomotive.) It was not successful, but its problems were not related to the simple-articulated concept. It was designed in a period when it was thought that superheaters allowed the use of lower boiler pressures because of the increased heat content of the steam, but that theory was found to be in error. Unlike the Mallet compound, on which the boiler fed only two high-pressure cylinders, the simple articulated needed a large and potent boiler to supply steam to all four cylinders, so the 160-pound boiler pressure of the HH1s severely limited the machine.

The PRR tried the simple-articulated concept again with the HC1s of 1919 (another class of one), using PRR's standard 205-pound boiler pressure. It was condemned as too powerful. It had a feature called limited cutoff. Following the example of PRR's I1s 2-10-0 of 1915, the HC1s had a limited cutoff of 50 percent. This meant that the locomotive could get into a rod position where one side of the engine was on dead center and the other was at the point of cutoff, so no power could be produced. So-called starting ports had to be incorporated into the cylinders of the I1s engines to move them out of this rod position, but PRR designers evidently figured that it was unlikely that both engines of a simple articulated would be so affected. Engineers starting trains with the HC1s often had to jockey the reverse lever back and forth and, when the engine's valves did admit steam, the engine could lurch ahead violently enough to break coupler knuckles. The HC1s wound up in pusher service on the mountain out of Altoona, Pa. The HH1s and HC1s undoubtedly helped sour the Pennsy's attitude toward articulated locomotives in general.

The simple articulated concept lay dormant until 1924 when American Locomotive Co. built a group of 2-8-8-2s for the Chesapeake & Ohio. These were fearsome-looking engines whose actual capabilities were not any greater than the USRA-designed 2-8-8-2 Mallets. But the C&O engines, classed H-7, were publicized as being successful, and the railroad obtained 45 of them.

Speed sells simples

The simple articulated grew up, and designers and pundits touted them as being superior to the Mallet compound because they were faster; speed was always an easy sell for

Significant articulated locomotives

Road	Type	Class	Nos.	Qty.	Builder	Date	Retired	Notes
AT&SF	2-6-6-2	1157	1157	1	Santa Fe	1910	1924	Jointed boiler
AT&SF	4-4-6-2	1398	1398-1399	2	Baldwin	1909	1915	Largest drivers on an articulated
AT&SF	2-10-10-2	3000	3000-3009	10	Baldwin/Santa Fe	1911	1915-1918	Rebuilt from and to 2-10-2s
B&O	0-6-6-0	DD-1	2400	1	Schenectady	1904	1938	America's first Mallet
B&O	2-6-6-2	KK-1	7400	1	Baldwin	1930	1953	
C&O	2-6-6-6	H-8	1600-1659	60	Lima	1941		Most powerful steam locomotive built
C&O	2-8-8-2	H-7	1540-1564	25	Alco	1923-1924	1952	First successful simple articulated
DM&IR	2-8-8-4	M-3	220-237	18	Baldwin	1941, 1943	1958-1963	
ERIE	2-8-8-8-2	P-1	5014-5016	3	Baldwin	1914, 1916	1929-1933	
N&W	0-8-8-0	X1	990-994	5	Schenectady	1910	1934	
N&W	2-6-6-2	Z1, Z1a	1300-1489	190	Alco	1910	1958	
N&W	2-6-6-4	A	1200-1242	43	Norfolk & Western	1936-1950	1958-1961	Built in four batches
N&W	2-8-8-2	Y1	995-999	5	Baldwin	1910	1924	
N&W	2-8-8-2	Y3	2000-2049	50	Alco and Baldwin	1919	1957-1958	USRA design
N&W	2-8-8-2	Y6	2120-2154	35	Norfolk & Western	1936-1940	1958-1960	
NP	2-8-8-4	Z-5	5000-5011	12	Alco and Baldwin	1928-1930		Largest steam locomotives at the time
PRR	2-8-8-2	HH1s	3396	1	Alco	1911	1928	First simple articulated
PRR	2-8-8-0	HC1s	3700	1	Pennsylvania	1919	1929	
SAL	2-6-6-4	R-1, R-2	2500-2509	10	Baldwin	1935-1937	1947	Sold to B&O
UP	4-6-6-4	CSA, 4664	3900-3999	105*	Schenectady	1936-1943	1962	*CSA class renumbered to 3800 series
UP	4-8-8-4	4884	4000-4024	25	Schenectady	1941, 1944	1959	"Big Boy"
VGN	2-6-6-6	AG	900-907	8	Lima	1945	1960	C&O H-8 copies
VGN	2-8-8-2	AD	601-606	6	Alco	1912	1934	
VGN	2-10-10-2	AE	800-809	10	Alco	1918	1948-1958	
VGN	2-8-8-8-2	XA	700	1	Baldwin	1916	1920	

AT&SF: Atchison, Topeka & Santa Fe; B&O: Baltimore & Ohio; C&O: Chesapeake & Ohio; DM&IR: Duluth, Missabe & Iron Range; ERIE: Erie; N&W: Norfolk & Western; NP: Northern Pacific; PRR: Pennsylvania; SAL: Seaboard Air Line; UP: Union Pacific; VGN: Virginian. Source: *Guide to North American Steam Locomotives, Revised Edition*

the steam locomotive salesman. The economics of compounding inherent in the Mallet went by the board as the builders decided that the simple articulated was better for them in two ways: First, simple articulateds had to have larger boilers to supply the steam, which meant that they would be more profitable for the builders to manufacture; and second, the builders didn't have to go to the trouble to identify the factors that kept the Mallet from being faster, and then do something about them.

Lots of simple articulated locomotives hit the road in the late 1920s and 1930s. Among these were the immense 2-8-8-4 Yellowstone-class for the Northern Pacific, which were the largest locomotives in the world at the time.

Several railroads, motivated by that search for speed, converted Mallet compounds into simple engines. The Baltimore & Ohio, Great Northern, Southern Pacific, and Southern Railway all replaced those large, low-pressure cylinders with cylinders the same size as the high-pressure ones, arranging the steam and exhaust piping appropriately, and then they bragged about how much faster and more efficient these locomotives were. On the B&O, these engines were used on heavy grades, both as road engines and as helpers. But how much extra speed was going to be useful in these applications? Tonnage limited speed going up and curvature limited speed going down, where the engines were drifting anyway. But still, the railroads implicitly insisted, nobody ever built or converted a lemon.

Another interesting facet of these rebuilds is that no USRA 2-8-8-2 on any railroad was ever converted to a simple engine. Evidently the USRA engines were fast enough that any extra speed wasn't worth sacrificing the economy of compounding.

Norfolk & Western, though, believed that there was speed potential in the USRA 2-8-8-2 that it could obtain without sacrificing the economy of the compound. As the rest of the world was accumulating rosters of the faster simple engines, N&W put its dynamometer car to work with its Y2a, Y3, and Y3a engines — 30 of which were obtained in 1923 — to identify areas where they could be modified to run faster. Boosts in boiler pressure were seen as beneficial; all the locomotives had their working pressure raised in increments to 270 pounds, a 12-percent increase. This made them both faster and more efficient.

The low-pressure cylinders of the Mallet caused problems. Steam came into the back of the low-pressure cylinder castings from the receiver in the middle of the low-pressure engine frame, and had to pass down beneath the frame and then up to the top of the casting to the valves. The exhaust-steam passages had to share a fairly limited space to go down

Virginian's 10 Alco-built 2-10-10-2s of 1918 featured 48-inch-diameter low-pressure cylinders, the largest cylinders ever applied to a steam locomotive. The well-dressed men provide the modern viewer a sense of scale. *Trains collection*

Norfolk & Western A Class 2-6-6-4 No. 1232 speeds a freight along the Portsmouth-Columbus, Ohio, main line in 1957. The road had main lines to both Columbus and Cincinnati. *Robert Hale*

The Pennsylvania's sole HH1s class 2-8-8-2 was the first simple articulated to ride the rails when it debuted in 1911. The "s" in the designation indicates a superheater-equipped locomotive, which was not able to make up for its paltry 160-pound boiler pressure. *Trains collection*

beneath the frame and up to the top center of the cylinder castings, from where it was piped through swiveling connections to the exhaust stand and the nozzle in the smokebox. N&W redesigned these low-pressure cylinders, making the steam passages as generous as possible, and ordered another 10 2-8-8-2s from Alco's Richmond Works in 1927. These Y4 engines were, indeed, found to be faster than the earlier models, but the railroad was still not satisfied. In 1929, Y3 No. 2049 was taken into Roanoke shop and emerged with a totally new concept for a low-pressure exhaust system. The exhaust was taken out of the top of the cylinder casting and the two sides were merged by what was termed a bridgepipe: From the bottom center of the bridgepipe an exhaust pipe in the shape of a shallow "u" connected it with the exhaust stand in the smokebox. No longer did the steam from the receiver have to share space in the cylinder casting with the exhaust passages. Dynamometer testing with the 2049 was gratifying, and it convinced the railroad that every Mallet it would obtain thereafter should have the bridgepipe.

As the 1930s began, Baltimore & Ohio bought from Baldwin the first high-wheeled simple articulateds — two 70-inch-drivered 2-6-6-2s, one with an Emerson water-tube firebox and the other conventional. The engine showed front-end stability problems at the

speeds promised by the high drivers. Alco, also interested in getting into the high-speed articulated locomotive market, noted these problems and decided to combat them by providing a four-wheel leading truck to smooth out operation.

Stump pullers

The N&W, having an expanding need for a brawny "stump puller," opened the 1930s by building 20 more 2-8-8-2s that incorporated all that the railroad had learned thus far. The new engines would have big nozzles, bridgepipes, and a boiler pressure of 280 pounds, later raised to 300. The Y5-class proved not only to be strong (a starting tractive effort of 152,206 pounds in simple, 126,838 in compound), it was a runner, too. But the Y5s had been built without two innovations in steam locomotive hardware — cast-bed frames and roller bearings — and as the 1930s wore on they were plagued with broken frames and hot driver journals.

The mid-1930s saw Baldwin build five 2-6-6-4 engines for the Seaboard Air Line with 69-inch drivers for fast freight and passenger service between Richmond, Va., and Hamlet, N.C. Baldwin had evidently learned some things about front-end stability since the B&O

An eastbound Northern Pacific freight assaults Montana's Bozeman Pass in July 1948 with Z-5 class 2-8-8-4 Yellowstone No. 5003 shoving behind wood caboose No. 1321. Duluth, Missabe & Iron Range also carried 18 of the wheel arrangement on its roster beginning in 1941. *Bob Milner*

engines of 1930. The Seaboard engines were regularly operated without incident at speeds of 60 mph and more. Built with a 55,000-pound axle loading, lighter than many 4-8-4s, they handled the tonnage of two modified USRA light 2-8-2s, which were excellent engines, and they did it faster. Their success encouraged Norfolk & Western to design its own 2-6-6-4, also a successful locomotive, and first built in May 1936.

The 2-6-6-4 relieved the N&W from having to build a super-fast stump puller, but the railroad decided to build five new 2-8-8-2s later in 1936 that would incorporate bed castings and roller bearings. The one-piece bed casting allowed the low-pressure cylinders, now part of the frame, to have ultra-generous steam passages from the receiver — itself an integral part of the low-pressure engine frame — to the valves at the top of the cylinder casting. Since the bridgepipe took the exhaust steam out of the top, steam flow through the low-pressure engine was as free as could possibly be imagined. The new engines, classed Y6, forecast the ultimate compound.

We have enumerated the methods used by Norfolk & Western to make the Mallet compound faster. These are not steps that any mechanical engineer couldn't have devised, properly motivated. Those working for the commercial locomotive builders were motivated by a need to help their firms maximize profits, mainly by producing designs that could be sold easily. They created a competitive environment based on a need for speed. In the 1920s, this necessity was due to competition from motor trucks, which had the advantage of being able to provide door-to-door service no matter where the shipper and receiver were located. The Super-Power concept took hold (see Chapter 6), and it sold well. Super-Power freight locomotives — principally 2-8-4s and 2-10-4s with 100-square-foot grates under deep fireboxes and riding on 69-inch drivers — made mile-a-minute freight trains a reality, and as noted above, speed was always an easy sell, and the motor truck made it an even easier sell.

There were a couple of examples that demonstrated just how far railroads, and locomotive designers, were willing to go to avoid the use of an articulated locomotive of either kind — Mallet compounds or simple articulateds. For the Erie, Baldwin diagrammed a 2-10-4 in 1929 with 77-inch drivers, 31-inch cylinder bore and 32-inch stroke, and a boiler pressure of 260 pounds. Those 2-10-4s were never built: Erie's highly competent fleet of 105 2-8-4s and the stock market crash put an end to the idea.

The Santa Fe had dabbled unsuccessfully with Mallets between 1909 and 1911: The experiments included 2-6-6-2s with jointed boilers, passenger 4-4-6-2s with 73-inch

AIR CONDITIONED ICE REFRIGERATORS
COAL AND OIL
2502

Seaboard Air Line 2-6-6-4 No. 2502 leads a freight at Raleigh, N.C., in 1941. Seaboard's 10 such locomotives predate Norfolk & Western's arguably more famous entries by a year. They operated in fast freight and passenger service north of Hamlet, N.C. *Trains collection*

drivers (the highest drivers ever on an articulated), and 2-10-10-2s made from conventional 2-10-2s. The Santa Fe was thus receptive to the idea that it might obtain articulated performance from a two-cylinder locomotive. The result was a class of enormous 2-10-4s. The 5001-class had cylinders with a 30-inch bore and 34-inch stroke, a boiler pressure of 310 pounds, riding on 74-inch drivers. With their 4-6-6-4-sized boilers, the engines weighed 545,000 pounds. It would be heresy not to consider these engines among the greats, but their tractive effort of 93,000 pounds was eclipsed by most 4-6-6-4s, and those immense drivers can't have been an asset on heavy grades. Looking at the 5001-class engines, and considering the weight of the main and side rods necessary to transmit that power, a question arises. Did the engines have 74-inch drivers because they were expected to operate in the 75-80 mph range of the similarly equipped 4-8-4s, or were the large wheels necessary to provide room for proper counterbalancing for those massive rods?

Norfolk & Western's locomotive designers weren't motivated by the necessity to sell locomotives. They were, however, motivated by the need to have their employer make as much money as possible. The normal parameters of gross ton-miles per train-hour and gross income carried over to net were N&W's yardsticks, but the railroad was handicapped by hauling a low-revenue commodity, coal, over a mountainous profile with heavy curvature. Its mountain freight power had to be compact, potent, and as economical to operate as possible. This is why the concept of compounding, using steam twice, had such great appeal.

Making the physics work

The steam locomotive was beset by physical drawbacks. One of them was that the locomotive needed wheels ahead of the drivers to make it stable as it moved ahead; another was the need for wheels behind the drivers to bear the weight of a large firebox. Out of a total of seven engine axles, a 4-6-4 Hudson-type only had three that produced power. (Some had trailing-truck boosters, but they weren't effective at running speeds.) On a 4-8-4 Northern-type, half the wheels were for guiding or carrying, and not for producing power at speed.

As far as articulated locomotives were concerned, the two most likely to be considered as the ultimate were the Union Pacific 4-8-8-4 Big Boys and the Chesapeake & Ohio and Virginian 2-6-6-6 Allegheny types. Big Boy carried 71.6 percent of its weight on its 16 drivers. The 2-6-6-6, the first of which outweighed the 4-8-8-4 by 3 tons, only had 65.2

604
604
604
K AND WESTERN
2147
2147

Norfolk & Western's three most famous steam locomotive types: left to right, Class J 4-8-4 No. 604, Class Y6 2-8-8-2 No. 2147, and Class A 2-6-6-4 No. 1212. Sister engines Nos. 611 and 1218 would rise to fame in the 1980s in successor Norfolk Southern's steam program.
Norfolk & Western

percent of its weight on its 12 drivers. It had more weight on its three-axle trailer than the total weight of many good-sized Consolidations.

But if one considers the four Baldwin-built Triplexes (the three Erie 2-8-8-8-2s built in 1914 and 1916, and Virginian's monumentally unsuccessful 2-8-8-8-4 of 1916) you get more productivity per pound of total engine weight — 93 percent in the case of the Virginian engine. Among more successful locomotives, Alco's 2-10-10-2s of 1918 for the Virginian had 90.2 percent of their engine weight available for adhesion.

The N&W's Y5, Y6, Y6a, and Y6b-class Mallets carried 89.7 percent of their weight on their 58-inch driving wheels. But all this came to naught before the onslaught of Dick Dilworth's EMD FT-model diesel No. 103, which carried 100 percent of its weight on its driving wheels. And, when ballasted to the axle-loading parameters of 55,000 pounds used in light USRA engines, it could produce at 25 percent adhesion that much tractive effort per unit, or 220,000 pounds for the four-unit consist. Dilworth didn't even have to produce a six-motor locomotive to out-lug steam's best at low speed — on the mountains, for instance.

Did Norfolk & Western get any payoff by sticking with compounding for its stump puller? Its operating and financial statistics from the 1950s say yes. In spite of its topography, N&W consistently ranked among the top railroads in gross ton-miles per train-hour. Its ratio of gross income carried over to net was also the envy of other roads.

But how did the 2100-series 2-8-8-2s actually do? Let's take, for instance, coal trains between Bluefield and Roanoke, Va. The standard train was about 130 cars, or 10,300 tons. These trains dropped down the mountain out of Bluefield and came out in the New River valley at Glen Lyn, Va. From Glen Lyn they came up New River against a 0.2-percent grade and constant curvature to Walton, Va., where they got another 2100-series pusher for the 1-percent grade for 7 miles to Christiansburg, Va., where the pusher cut off on the fly.

There's a mountain of anecdotal evidence that tells that the 2100-class engines came up New River at speeds between 30 and 32 mph. This topography and train consists have been run through the train-resistance formula, which indicates that 5,500 hp was required on the drawbar of the tender. How the 2100s did it was by getting far more work out of the low-pressure engine at speed than any other Mallet ever dreamed of; instead of being a liability at speed, the big low-pressure cylinders were doing their share of the pulling.

Were there other steam locomotives that could pull such a train up New River this fast? Certainly UP's Big Boy; Lima's Allegheny-types for C&O and Virginian; probably the

Duluth, Missabe & Iron Range M-3 class 2-8-8-4 No. 228 departs Proctor, Minn., westbound with 200 empty ore cars under a column of exhaust in August 1959.
William D. Middleton

228
228
DULUTH MISSABE
& IRON RANGE
SAFETY
DULUTH
MISSABE
&
IRON RANGE
FIRST
CAP 140000

Duluth, Missabe & Iron Range 2-8-8-4 Yellowstone-types; and Great Northern's simple 2-8-8-2 could have done it. But where the N&W's 2100 aced them all was in doing it with a boiler comparable to a good-sized 4-8-4. All the rest had humongous boilers with appetites to match. The efficient use of steam twice is what did the trick. And it's likely that it would have taken more than two of the others to get the train up Alleghany Mountain to Christiansburg.

What about those lemons?

The world of articulated locomotives also has its lemons. First was the "centipede" tender, which was one answer to getting seven axles under a high-capacity tender. Centipedes were impressive, but the five-axle rigid wheelbase didn't lend itself to reverse moves on sharp curves. The Clinchfield, which spun its locomotives on wye tracks at several locations instead of turntables, found that its six UP/Alco-design 4-6-6-4 Challengers wanted to straighten out wye tracks.

Lima then was faced with the need to put seven axles under the tenders for the H-8 class 2-6-6-6s it was building for the Chesapeake & Ohio. Ultimately, Lima put a three-axle truck forward under the coal space and a four-axle truck behind under the water end — not as impressive as the centipedes but more practical.

Another articulated lemon was the aforementioned C&O 2-6-6-6. It was not designed to be an efficient source of power for C&O, but rather to develop the most drawbar horsepower of any steam locomotive. This it did, in spectacular fashion, but at tremendous cost. The engines cost an average of $100,000 per engine more than N&W's 2-6-6-4, and outweighed the N&W engine by a full hundred tons. The C&O had a 160-car limit on coal trains between Russell, Ky., and Toledo, Ohio. Its T-1 2-10-4s could handle such a train, but we have been told by all that we must believe that the Allegheny was one of the all-time greats. Perhaps, on the C&O at least, less was more.

ED KING JR. is retired from a long career in railroading, including stints with Norfolk & Western, Rock Island, Soo Line, and others. He has more than 30 TRAINS bylines and is a former columnist for the magazine. This article appeared in Big Boy: Back in Steam, *a TRAINS special edition, in 2019.*

A Norfolk & Western coal train eases over the long bridge west of Cooper Tunnel near Bluefield, W. Va. The author contends that the railroad's use of compound locomotives was a great benefit to its operations in difficult mountain topography. *O. Winston Link*

5 Which Was the Biggest?

by Ed King Jr.

Comparing Big Boy to America's other heavyweight contenders

THE ONSET OF WORLD WAR II PRODUCED two of railroading's most impressive steam locomotives — Chesapeake & Ohio's Lima-built 2-6-6-6 and Union Pacific's Alco-made Big Boy 4-8-8-4.

Everything about the Big Boy grabbed headlines — from its size (UP called it the "World's Largest Locomotive," although C&O's 2-6-6-6 was heavier and had about the same size boiler) to the appearance of the name "Big Boy" in chalk on the first engine, a happenstance milked heavily in advertising by Alco and UP.

Union Pacific historians are fond of saying that the Big Boy could have produced even more than its 6,300 drawbar horsepower had it used better coal. But that's difficult to understand, given that during its tests the Big Boy was operated at full capacity and the boiler was fully supplying the demands of the machinery. If higher quality fuel had been used, it might have been able to use less of it. But why design one of the all-time ultimate steam locomotives and handicap it by using inferior fuel? If high-quality coal was expensive to come by, why not burn oil?

A curious feature of Big Boy was the use of an exhaust steam injector instead of a more efficient and reliable Worthington or Elesco feedwater heater. Exhaust steam injectors were called "poor man's feedwater heaters," but UP was not a poor man. Big Boy historians say that rather than deal with the tricky starting sequence of the exhaust steam injectors, many engine crews avoided their use and ran on the regular injector alone, which defeated the purpose of the application. That seems strange in light of UP's carefully nurtured reputation for requiring only the best of everything.

Alco and Union Pacific also promoted the claim that Big Boy's machinery was designed for 80-mph operation. But USRA engines of 1919 — the 4-8-2s, specifically — with drivers only an inch larger than Big Boy's, and counterbalanced with 1919 technology, had been capable of running 80 mph for 22 years. And other than attracting publicity, what was the actual value of running Big Boys at 80 mph? How much of Big Boy's running time was spent going more than even 65 mph?

What Union Pacific needed was a locomotive capable of handling tonnage up the Wasatch and Sherman grades and then running 60 mph between those summits. With Big Boy, the railroad got more speed than it could make use of, and less tonnage-hauling capacity than it should have, given the engine's weight. They sure were pretty, though.

The C&O and Lima got similar publicity for the 2-6-6-6. The Allegheny proves that not all Super Power was created equal. It is well documented that the class H-8 2-6-6-6 had been designed to outperform N&W's class A 2-6-6-4 of 1936. The Advisory Mechanical Committee of the C&O and other Van Sweringen roads had more than five years to consider the performance of Seaboard's 2-6-6-4s of 1935 and UP's Challengers of 1936 to aid in designing a new high-speed articulated. The committee knew that C&O would buy anything it recommended and was willing to spend as much of C&O's money as necessary to accomplish its goal.

Norfolk & Western's A had produced 6,300 drawbar horsepower in testing, and C&O and Lima people on their own dynamometer car shed tears of joy when that figure was exceeded by the Allegheny to a maximum of 7,498 drawbar hp. But the first 10 class A

Chesapeake & Ohio Allegheny-type 2-6-6-6 No. 1622 pulls a perishables train at Alderson, W.Va., in the 1940s. The locomotive was heavier than Big Boy, but with less weight on drivers. *B.F. Cutler*

Union Pacific Big Boy 4-8-8-4 No. 4017 storms Sherman Hill at Emkay, Wyo., in the 1950s. *William W. Kratville*

Home-built Norfolk & Western 2-8-8-2 Y6b No. 2173, built in 1948, leads an eastbound coal train at Blue Ridge, Va., 1958. Sister Y6b No. 2200 became the last road freight steam locomotive built in the U.S., emerging from Roanoke in April 1952. *Jim Shaughnessy*

Northern Pacific class Z-5 2-8-8-4 No. 5006 poses at Livingston, Mont., 1954. Baldwin built 11 of the Yellowstones for NP in 1930. The Z-5 had the largest grate area of any locomotive (182 square feet) for burning low-grade Rosebud coal. *Harold K. Vollrath*

5006

Duluth, Missabe & Iron Range class M-4 2-8-8-4 No. 232 came from Baldwin in 1943. The DM&IR's Yellowstones carried ore trains on the railroad's Missabe Division.
Starlake Films

engines weighed 570,000 pounds vs. the Allegheny's 778,000 (the heaviest reciprocating steam locomotives ever built). Each of the first 10 N&W 2-6-6-4s cost $123,448, while C&O paid $230,663 for each of the 10 earliest 2-6-6-6s (this figure is not out of line; UP paid Alco $133,000 each for its 1936 4-6-6-4s. The N&W's Roanoke Shops, of course, didn't have to earn a profit). The 2-6-6-6 carried nearly as much weight on its trailing truck as the total engine weight of a Southern class Ks 2-8-0, and N&W's entire class M 4-8-0 weighed only about three tons more.

So the Allegheny did, indeed, beat the A's horsepower. But for an additional 100 tons of weight and an extra $100,000 per engine and tender, shouldn't it have? You can do the math and form your own conclusion about whether C&O got its money's worth out of all that additional weight and expense.

The 2-6-6-6 is today considered by many to be the ultimate expression of the Super Power concept. However, C&O put these engines to work in the mountains, where they seldom, if ever, got to use all that horsepower, and their 110,200 pounds of starting tractive effort limited the tonnage they could handle. Their rating was only 200 tons more than the H7 2-8-8-2 of the 1920s. The 2-6-6-6 fared better hauling tonnage up through Ohio on C&O's Russell (Ky.)-Toledo run, but because of a car limit imposed by the operating department, Chessie's T-1 2-10-4s (probably Lima's finest locomotive) of 1930 could do the same job much more economically. But what happened to the idea that the engine should be a tool to make as much money as possible for its owner? Was it a sound business decision to design and build such an expensive and heavy locomotive to outperform another locomotive? Rather than admit a mistake, C&O went out and bought 60 of them.

Virginian 2-8-8-8-4 No. 700 posed for its Baldwin builder's photo in 1916. Triplexes had their third engine under the tender. They tended to use up steam too quickly. The locomotive was not successful; it was sent back to Baldwin in 1920. *Baldwin Locomotive Works*

Virginian 2-10-10-2 No. 801, at Victoria, Va., was one of 10 built by Alco in 1918. It offered a boiler pressure of 215 psi and led all steam locomotives in tractive effort (147,200 pounds as a compound; 176,600 pounds simple).
H.P. Stearns

What was the biggest in steam? Take your choice

Railroad	Series	Wheel arrangement	Boiler pressure (pounds/sq. in.)	Drivers (inches)	Weight on drivers (pounds)	Engine weight (pounds)	Engine & tender weight (pounds)
Chesapeake & Ohio	1600-1659	2-6-6-6	260	67	498,000	778,000	1,208,000
Great Northern	2044-2059	2-8-8-2	240	63	544,000	630,750	1,003,530
Norfolk & Western	2171-2200	2-8-8-2	300	58	548,500	582,900	990,120
Duluth, Missabe & Iron Range	228-237	2-8-8-4	240	63	565,000	699,700	1,138,000
Northern Pacific	5001-5011	2-8-8-4	250	63	558,900	723,400	1,125,400
Union Pacific	4000-4024	4-8-8-4	300	68	545,200	772,250	1,208,750
Erie	5014-5016	2-8-8-8-2	210	63	761,000	753,000	845,000
Virginian	700	2-8-8-8-4	215	56	726,000	844,000	844,000
Virginian	800-809	2-10-10-2	215	56	617,000	617,000	684,000
Notes: S=simple. C=compound. B=with booster.							

Erie's three Baldwin 2-8-8-8-2 triplex locomotives, including No. 5016, spent their lives in pusher service. They were retired in 1927. *Trains collection*

	Cylinders: diameter x stroke (inches)	Evaporative surface (square feet)	Grate area (square feet)	Firebox, length & width (inches)	Tractive force (pounds)	Builder and date
	$22\frac{1}{2}$ x 33	7,240	135	180 x $108\frac{1}{4}$	110,200	Lima 1941, 1948
	28 x 32	7,947	126	168 x 108	146,000	GN 1929-1930
	25 x 32 & 39 x 32	5,656	106.2	$143\frac{3}{4}$ x $106\frac{1}{4}$	132,000 (C), 170,000 (S)	N&W 1948-1952
	26 x 32	6,758	125	$210\frac{1}{8}$ x $102\frac{1}{4}$	140,000	Baldwin 1943
	26 x 32	7,666	182	$266\frac{1}{4}$ x $114\frac{1}{4}$	140,000, 153,300 (B)	Baldwin 1930
	$23\frac{3}{4}$ x 32	5,755	150.3	$235\frac{1}{32}$ x $96\frac{3}{16}$	135,375	Alco 1941/1944
	36 x 32	6,886	90	162 x 108	160,000	Baldwin 1914
	34 x 32	8,120	108.2	188 x $108\frac{1}{4}$	146,000	Baldwin 1916
	30 x 32 & 48 x 32	8,606	108.7	$181\frac{1}{16}$ x $108\frac{1}{4}$	147,200 (C), 176,000 (S)	Alco 1918

Built in Great Northern's own Hillyard (Spokane) shops, class R-2 2-8-8-2 No. 2055 weighed more than 1 million pounds. The railroad built 16 of them in 1929 and 1930.
Trains collection

Virginian 2-6-6-6 No. 902 at Norfolk, Va., was a copy of C&O's Allegheny design that was built after a C&O executive moved to the line. Lima built 8 for the railroad in 1945; they served until 1960. *H. Reid*

The wartime president of C&O neighbor Virginian, Frank Beale, came from C&O, where he'd been an operating official on Allegheny Mountain when the 2-6-6-6s came. Beale was impressed. He must have figured that the Virginian had no mountains east of Roanoke, so the 2-6-6-6 would work just fine. He bought eight, plus five C&O-style 2-8-4s for fast freights. For a 35-mph railroad, here were 13 big, heavy, expensive locomotives, each of which developed maximum drawbar horsepower at better than 40 mph. Like their C&O counterparts, the fans and historians liked them. The stockholders didn't have a choice.

This article appeared in Big Boy: On the Road to Restoration, *a* TRAINS *special edition, in 2014.*

6 Super-Power

by Neil L. Carlson

How modern steam technology evolved
from Berkshire to Big Boy

DURING THE 1920S, NORTH AMERICAN RAILROADS found themselves in the difficult position of not having enough capacity. They were straining to handle existing traffic, much less accommodate the growth generated by a booming economy. After the extreme traffic conditions of World War I, important routes still lacked modern signal systems. Track suffered from deferred maintenance, rights-of-way were under-engineered, and substantial bottlenecks existed. Furthermore, many railroad rosters included a lot of locomotives that by 1920 were obsolete.

Although the railroads clearly needed to modernize their physical plants, the easiest way to increase capacity was to run longer, faster trains. Ultimately, this is what happened, but it didn't occur overnight. First, the railroads had to overcome their "tonnage" mentality. Prior development had produced increasingly larger locomotives that had allowed train size to gradually increase, but speed was not in the equation. Nonetheless, as the railroads took stock of the situation, a demand was created for larger, faster, more powerful locomotives. This situation was not lost on the locomotive builders.

The Super-Power concept

On December 20, 1926, in a paper presented to the Western Railway Club in Chicago, William E. Woodard of Lima Locomotive Works made the point that while mainline mileage had not materially increased in the last 20 years, the traffic carried had grown enormously. The most important factors in handling this increased load were the advances in steam locomotive design that allowed heavier and faster trains.

Just the same, the steam locomotive was reaching the end of any performance improvements that could be expected with existing technology. Further development would be required, and this is what spawned the "Super-Power" era. It carried locomotive technology to a much higher level and was characterized by a series of innovations that permitted fewer—but bigger—locomotives to haul ever larger and faster trains. Super-Power also made important strides in improved locomotive availability and utilization.

Super-Power technology built upon developments that had greatly advanced locomotive design in the prior 20 years. To understand what this means, let's go back and look at the improvements that took place during this period.

For the first 75 years of the steam locomotive, it was not possible to design a locomotive based on scientific principles — the study of thermodynamics was still in its infancy. Instead, designers relied on intuition — and trial and error. New designs were often laid out by choosing an existing locomotive and "slide-ruling" its dimensions up or down to create a new design, and many locomotives were built through use of this rather crude technique. Some were winners, such as the 4-4-0. By 1850 it had become *the* standard locomotive, and it proved to be a well-balanced design — it had such natural proportions that it was difficult to build a bad one. The firebox, located between the driving wheels,

In 1925, Lima's groundbreaking A-1 2-8-4 debuted the Super-Power concept. When No. 1 tested on the Boston & Albany, the housing on the pilot deck shielded technicians. The wheel arrangement earned the name Berkshire following its time on the B&A. *Lima Locomotive Works*

The first of the Big Boys, No. 4000, thunders past the depot at Sherman, Wyo., with an eastbound freight in August 1947. Sherman, with an elevation of 8,015 feet, is the highest point on the original Union Pacific system. *R.H. Kindig*

The group of vertical pipes in this smokebox is the header of a type A superheater. With a type A, steam passes four times each through a limited number of enlarged flues. In the later type E, steam made two passes through each of a greater number of flues. *Bruce R. Meyer*

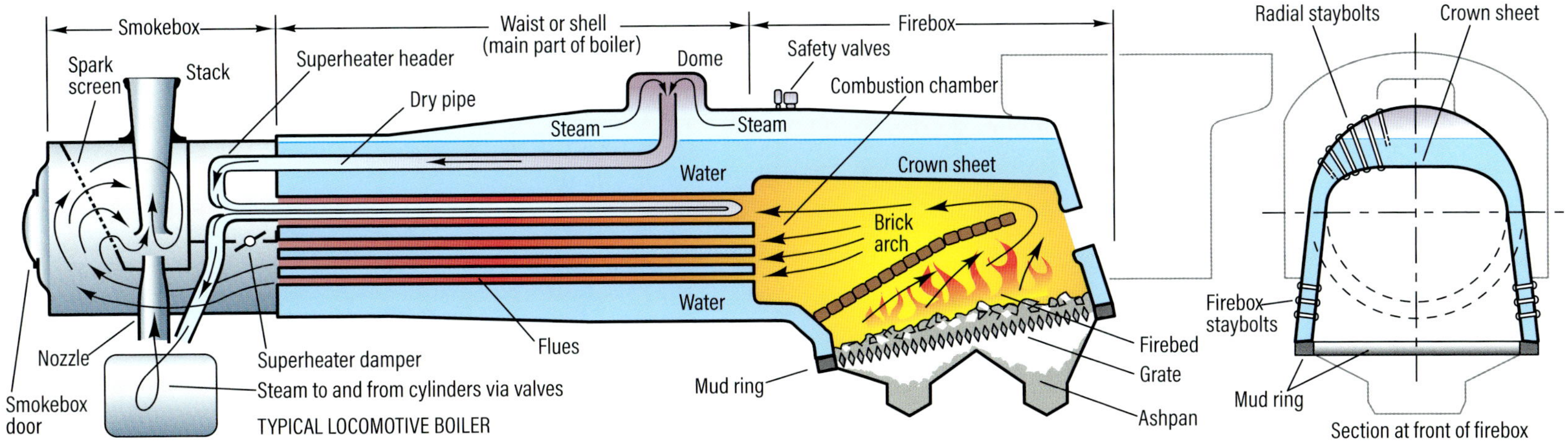

was deep and had ample volume for good combustion. There was also a good balance among grate area, heating surface, and cylinder volume.

During the 1880s, a degree of specialization took place between passenger and freight power, and new wheel arrangements were developed. The 2-8-0 became the standard for freight service, and from it came the larger 2-10-0. The 4-6-0 grew out of the 4-4-0 for heavy passenger and fast freight service. These new types did not have the natural proportions of the 4-4-0, however, and some poor performers were built.

In an attempt to provide design direction, the American Railway Master Mechanics Association (ARMMA) in 1897 published a set of recommended ratios relating grate area, heating surface, and cylinder volume. However, these ratios offered only minimum values and provided no method for directly calculating the evaporation expected from a boiler. This was a serious limitation, since without it boiler horsepower could not be estimated. So, comparative data of other existing engines still had to be employed along with the ARMMA ratios. These ratios were an improvement, but some colossal flops continued to be built (several early Mallets equipped with fireboxes barely appropriate for 2-8-0's come to mind). A better system was still needed.

In 1912 science and locomotive design began to converge when Professor W. F. M. Goss of the University of Illinois conducted a series of boiler evaporation tests in Coatesville, Pa. The boilers used were segmented internally so that the water surrounding the firebox was separated from the water surrounding the flues. By this means, the amount of steam evaporated in each section could be separately measured and the amount of heat absorbed could be determined. Designers finally had a means to estimate boiler evaporation and therefore horsepower, and one could now predictably design a boiler to produce the desired output.

In 1914 Francis J. Cole of the American Locomotive Company (Alco) published a paper based on the results of Goss's Coatesville tests. Cole specified a series of practical design ratios which provided a means to directly calculate boiler and cylinder horsepower by using specifications from the boiler, firebox, and cylinders. A draftsman could now sit down and, by using the Cole ratios, reliably develop a new locomotive from scratch. He would start with the service requirements and then pick the cylinder dimensions that produced the required horsepower. Next, using the design ratios, he could proportion the boiler to provide the required heating surface and grate area. The Cole ratios, with some later modification to account for fire intensity, were extremely significant, for they were used by designers right up to the end of steam-locomotive production.

Superheating

A key concept was to build as powerful a boiler as circumstances would permit. The boiler was the prime mover, and if it could not turn water to steam at a rate sufficient for a locomotive's service requirements, it didn't matter how well-designed the rest of the engine might be. Without question, the single greatest advance in increasing boiler capacity came with the development of the superheater.

Superheating solved the condensation problem. To understand this, let's look at an example. The boiling temperature of water rises under pressure. For instance, at a pressure of 200 psi, water will begin to boil at 387 degrees F. To convert water to steam, it must absorb a specific amount of heat (the "heat of vaporization") — just to evaporate it. The temperature of the steam remains 387 degrees and the pressure remains 200 psi. Steam and water co-exist at this temperature and pressure. The amount of heat absorbed during evaporation is substantial — significantly more than it took to just raise the water temperature to 387 degrees.

This is called saturated steam. If any heat is removed from it, it will begin to condense back to water. When saturated steam performs work by expanding against a piston, some of its internal heat energy is converted into work, and as a result, steam will begin to condense. With a "simple" locomotive (see Chapter 4), as much as 20 to 25 percent of the total steam supplied to the cylinders condensed. On a compound locomotive it was even more. When saturated steam expands and condenses, the effective pressure of the remaining steam in the cylinder drops rapidly, and a locomotive loses power.

Superheating is the term for heating saturated steam above the temperature at which it was evaporated from water. When it is removed from direct contact with water in the boiler, the temperature of saturated steam will rise if more heat is added. In this example, if the steam were heated to 587 degrees, it would be said to have 200 degrees of superheat. When superheated steam performs work through expansion, it also gives up heat. But instead of condensing, only its temperature drops, and the effective pressure of steam expanding in the cylinder is not further reduced by condensation. This is a tremendous advantage over saturated steam.

Firebox design

Bituminous coal contains 25 to 50 percent volatile matter (hydrocarbons), with the remainder composed of fixed carbon, moisture, and various impurities. As coal burns, the volatile matter is driven off in gaseous form, and after mixing with air, it burns above the fuel bed in the volume of the furnace. Fixed carbon can combine directly with air and burn on the grate. If sufficient air is lacking for complete combustion, it might just partial-

Union Pacific's tinkering with front-end designs is seen in 4-8-4 No. 844's twin exhaust stacks, ahead of which is a rectangular Worthington SA feedwater heater. *Ben Fredericks*

ly burn, resulting in the creation of carbon monoxide. This is a highly flammable gas that also burns above the fuel bed. Sufficient furnace volume is a critical requirement for these gases to be mixed with air and completely burned.

In locomotives, the necessary firebox volume grew as the firing rate increased, since more gases were driven off and more space was required for complete combustion. When an engine was working hard, these gases moved very rapidly, and if they were not fully burned by the time they entered the boiler flues, they went up the smokestack as wasted energy. Firebox volume needed to be in correct proportion to grate area. Modern practice called for a firebox volume in cubic feet of about 6 to 7 times the grate area in square feet.

Fuel oil is virtually all volatile matter. In locomotives, it was atomized as it was sprayed into the firebox. It mixed with air, and then was burned. Just as with a coal-burner, sufficient firebox volume was an important consideration in an oil-burning locomotive.

Early locomotive fireboxes rested between or on the locomotive frames, in either case between the driving wheels. Fireboxes of this era were narrow and deep. Locomotives of the period were all hand-fired, and the deep firebox provided ample volume for good combustion given the low firing rates. As locomotives grew in size, larger grates were needed, and the grate was moved to be above the drivers. This permitted grates as wide as the locomotive, but it also created a shallow firebox with less volume relative to the grate area.

Fireboxes were later improved with the addition of the brick arch. This created a longer path for flame travel by forcing the burning gases to the rear of the firebox, up around the arch, and forward to the flues. It improved combustion by causing better mixing of air and fuel gases and soon became a standard feature on all coal-burning locomotives — but not oil-burners. The Santa Fe ran tests in 1915 with 2-10-2s equipped with and without brick arches. The results showed that the arch did nothing to improve combustion in an oil-burner. After these trials, most oil-burners were built without an arch, and coal-burners converted to oil-firing usually had the arch removed.

As more powerful locomotives were built, the large grates employed ultimately reached the limit of hand firing, and mechanical stokers were applied. But while the stoker solved one problem, it created a new one in that it could overtax the firebox. With the shallow fireboxes then in general use, there was just not enough volume to completely burn all the fuel that the stoker was capable of putting on the grates. As a consequence, economy and performance suffered.

The innovation of the trailing truck provided a solution. It could support a wide and deep firebox located below and behind the drivers. This provided the elements needed for good combustion — even at relatively high firing rates — and locomotive power increased substantially. With the trailing truck, the 2-8-2 Mikado and the 4-6-2 Pacific became the standards for freight and passenger service, respectively, and the 2-10-0 Decapod grew into the 2-10-2 Santa Fe for heavy freight service.

Feedwater heaters

The biggest efficiency loss in a locomotive comes from the tremendous amount of wasted heat in the exhaust steam which goes up the stack. Recall from the discussion on superheating that steam exhausted from the cylinders of a superheated locomotive does not condense. It still contains all of the specific heat used to convert it from water to steam. Further, the greater percentage of the heat from the fuel was used just in converting water to steam and not in superheating it. A feedwater heater recaptures some of this heat by using a portion of the exhaust steam to preheat boiler feedwater. It increases boiler horsepower and/or fuel efficiency by 8 to 10 percent. It also saves a similar amount of water as the condensed exhaust steam is recycled to the locomotive boiler. The benefits of feedwater heaters were so significant that in the Super-Power era almost all modern engines had them.

Feedwater heaters came in several forms. The most common systems utilized a separate heat exchanger — either of a closed or open design — and steam-driven pumps to force the feedwater into the boiler. In closed systems, exhaust steam was piped through a tank of feedwater, but it did not come in direct contact with the water. Elesco and Coffin feedwater heaters were of the closed type. The exhaust steam condensed to water as it passed through the heater, and it was piped back to the tender. In an open heater, the feedwater was sprayed directly into a mixing chamber filled with exhaust steam. The resulting heated water and condensed steam were then forced into the boiler. Ultimately, the open systems made by Worthington proved to be the most popular, and they became common appliances on many modern locomotives.

Lastly there was the exhaust steam injector — the so called "poor man's" feedwater heater. The injector was designed to use exhaust steam to both heat the feedwater and, supplemented by live steam, force it into the boiler. The Big Boy, incidentally, used an exhaust steam injector instead of a feedwater heater.

Front-end design

Exhaust steam that is ejected out the stack, while a major source of inefficiency, serves a necessary function. It induces a partial vacuum in the smokebox which pulls air via the boiler tubes through the grates to mix with the fuel in order to support combustion in the firebox. The hot fire gases are drawn through the tubes, and after transferring their heat to the boiler, are ejected up the stack.

Steam exhausted from the cylinders is led into a nozzle located beneath the stack, where it is directed upward. As it moves upward, it entrains the flue gases and carries them along with it. The ejection process is enhanced by both the speed of the exhaust steam and its surface area that is in contact with the flue gases. This is what induces the partial vacuum. As an engine works harder, more air needs to be drawn into the firebox to support the higher level of combustion. This happens automatically since the engine also is exhausting faster and more heavily, thereby increasing the draft.

However, heavier exhausts create a problem when the volume of exhaust steam becomes impeded by the nozzle. This creates back pressure in the cylinders which the pistons have to work against during each exhaust stroke. While some back pressure is necessary to induce a draft, too much is another source of inefficiency. Unfortunately, locomotive drafting was something for which there was no good theory.

The best front ends had exhaust nozzles that maximized the perimeter of the nozzle opening. This in turn maximized the surface area of the exhaust steam exiting the nozzle. As it turned out, the very common circular shaped nozzle was the least-efficient design. If you can picture a square nozzle with the same area as the circle, it will be evident that it has a greater perimeter than the circle. If you now do the same exercise with a rectangle, you can see that it has a still bigger perimeter than the square, and so on.

Through the use of a nozzle with a large area and a large perimeter, designers were able to keep back pressure under control and still have a good-steaming locomotive. Even though the exhaust steam did not have as high a velocity as would be the case with a smaller nozzle, its increased surface area more than made up for it. Many railroads developed different nozzle shapes based on this theme.

Perhaps no railroad tinkered more with front-end designs than the Union Pacific. It developed a system using double smoke stacks with a four-barrel "pepperbox" nozzle positioned beneath each stack. Santa Fe used the Layden nozzle that was its version of a four-barrel pepperbox. Norfolk & Western adopted an annular ported design after a series of successful tests (over 40 percent improvement in draft efficiency) using one of its K1 4-8-2s. The New York Central, Nickel Plate Road, and Chesapeake & Ohio were users of the basket bridge. All these designs were effective in forming the exhaust with more surface area to eject the flue gases.

Michigan Central 8000

For railroads with a "high-tonnage mentality," tractive effort was normally the most important locomotive specification. But as service requirements changed because of economic and competitive pressures, these carriers had to acquire locomotives capable of delivering power at speed. Horsepower then became as important as tractive effort, and it became incumbent upon the builders to develop the technology required to build more-powerful locomotives.

Among builders, Lima was at the forefront of this movement. Even before the first Super-Power locomotives, Lima was setting the stage with a heavy 2-8-2 designed with

Michigan Central (New York Central System) 2-8-2 Mikado No. 8000, harbinger of the Super-Power era, emerged from Lima in 1922 and quickly proved itself more efficient and more powerful than earlier locomotives. *Lima Locomotive Works*

horsepower uppermost in mind. However, it did not use all of the technology generally attributed to the Super-Power era. Built in 1922 as Michigan Central No. 8000, it was compared in a series of trials with a two-year-old, similarly sized heavy Mikado (class H-7e) of standard New York Central (MC's parent) design. The 8000 produced some spectacular results. It bettered its competitor by producing 15 to 28 percent greater drawbar horsepower at all speeds above 15 mph, while consuming no more coal and using less water.

As these locomotives were almost the same size, how were these results possible? To start with, No. 8000 had a bigger firebox. While it was the same length as that of the H-7e, it was wider, giving it a larger grate area. Furnace volume was also greater. The advantage of the larger grate and furnace volume was that the 8000 could burn more coal than the H-7e at any given firing rate (pounds of coal per square foot of grate per hour). Or, it could burn the same amount of coal per hour as the H-7e, but on the larger grate, it could be burned more slowly and efficiently.

The 8000 had a modern type E superheater unlike the H-7e, which had the older type A. The type E had several advantages. First, it produced steam at higher levels of superheat — as much as 50 to 100 degrees F greater. Second, because of the arrangement of tubes and flues made possible by its use, the boiler had a greater net gas area, a very important factor. This is the cross-sectional area of the flues through which the fire gases must pass as they go through the boiler from the firebox to the smokebox. When an engine is working, these gases can be moving at speeds in excess of 200 mph, so one can readily appreciate that this area has to be big enough to ensure that the gas flow will not be constrained. If the area is too small, an engine has to work harder to create sufficient draft to pull the gases through the tubes. This in turn means a higher back pressure and all the attendant inefficiencies that go with it. In the case of the 8000, the increased gas area obtained with

Illinois Central 7000 was a virtual copy of the A-1. Like other early Berkshires it had low (63") drivers. *Lima Locomotive Works*

the type E superheater nicely complemented the bigger grate. The increased grate area and net gas area taken together would account for 10 percent of the difference in performance over the H-7e.

Continuing with the boiler, we note that the 8000 had a pressure of 210 psi, 10 psi higher than that of the H-7e. This, along with the higher steam temperatures available with the type E superheater, would also improve performance. The increase in power due to the higher pressure and more superheat was about 6 percent.

Next, the 8000 was equipped with an Elesco bundle-type feedwater heater, the common design during this period. It is a closed-type design, in which the exhaust steam transfers its heat to the feedwater without coming in direct contact with the water. This feedwater heater would account for another 8 percent of the performance difference.

The last improvement was a subtle one. The 8000's auxiliaries operated on superheated steam instead of saturated steam. For the same reason that a locomotive operates more efficiently on superheated steam, so does the air pump, stoker engine, feedwater pump, turbogenerator, etc. This adds perhaps another 2 percent to performance as the auxiliaries now consume less steam.

Altogether, these improvements found on No. 8000 could account for a performance difference of 26 percent over the H-7e. But, recall that these improvements represent pre-Super-Power-era technology. They could just as easily have been applied to the H-7e. The real difference was Lima's design philosophy, which had horsepower in mind.

The New York Central was quite impressed by No. 8000 and ultimately bought 301 copies that became its class H-10. However, as good as they were, Will Woodard and his

Chesapeake & Ohio T-1 2-10-4s (Texas type), such as No. 3032, had 69" drivers, making them suitable for fast freight service. *W.G. Fancher*

The towering 74" drivers of Santa Fe's Texas types were second to none among 2-10-4s. *Baldwin Locomotive Works*

designers at Lima were already working on a successor, an engine capable of even greater performance.

The Berkshire revolution

Lima's A-1, the first 2-8-4 locomotive, ran a series of trials on the Boston & Albany in early 1925, and Lima's Woodard proudly discussed them in a December 1926 presentation to the Western Railway Club at Chicago. Built in 1924, the A-1 was unusual in that it was a demonstrator built by Lima on speculation. It had not been ordered by a railroad, as was the norm in the steam era.

In the series of B&A trials, the A-1 was compared with two-year-old B&A No. 190, an NYC H-10 class 2-8-2 (described earlier). The A-1 trials were conducted between Selkirk, N.Y., near Albany, and Washington, Mass. This 60-mile segment through the Berkshire Hills in western Massachusetts had many curves, with an average grade of 1 percent and a stretch of 1.5 percent approaching the summit near Washington. This line was operating at capacity, and B&A officials were interested in anything that would improve the situation. The A-1 made nine test runs between March 28 and April 18.

The results were spectacular, as the A-1 bettered the drawbar horsepower of No. 190 by 26 to 30 percent at all speeds. The B&A was extremely pleased with the results and ordered 55 A-1 copies, and thus was the 2-8-4 type given the name Berkshire. After leaving the B&A, the A-1 demonstrated on other railroads, and Lima wound up selling copies to Boston & Maine, Illinois Central, and Missouri Pacific. By the end of 1930, 156 locomotives of essentially the A-1 design were in service on the four railroads.

Four-wheel trailing trucks

The primary reason for the A-1's outstanding performance was its superior ability to evaporate water. Grate size had always been a limiting factor in boiler design because of the axle loading of the two-wheel trailing truck. This axle helped support the firebox, the grate, and the ash pan. On modern locomotives it also helped support a mechanical stoker, and increasingly, a booster engine. So, it was natural that when designers sought to increase locomotive power, they turned to bigger grates, and this required development of a four-wheel trailing truck.

The A-1's four-wheel truck permitted a firebox with 100 square feet of grate area, vs. 66.4 for B&A 2-8-2 No. 190. This allowed the A-1 to burn coal more efficiently, since it could burn it more slowly over a larger grate. Woodard cited an example of this: Measurements had been made while both locomotives were producing 1,600 drawbar horsepower. Engine 190 required 7,100 pounds of dry coal per hour, which equated to 107 pounds of coal per square foot of grate per hour.

The A-1 2-8-4, on the other hand, required only 4,750 pounds of dry coal per hour, or just 47 pounds per square foot per hour. This represented a 33 percent savings in coal. On the other hand, during periods when a high output was needed, the A-1's larger grate also permitted it to burn more coal than 2-8-2 No. 190. This liberated more heat for steam generation, and it increased the locomotive's output. After its introduction on the A-1, the four-wheel trailing truck became a trademark of Super-Power steam.

Norfolk & Western's home-built 2-6-6-4s, built beginning in 1936, had tall drivers and roller bearings. They were superb performers. *Richard J. Cook*

Larger drivers = power and speed

Unlike later Berkshires, the A-1 was not a fast freight engine. It had 63" drivers, while drivers of at least 69" were seen as a requirement for high-speed freight service. But in the late 1920s, no 2-8-4s (or 2-10-4s, a type introduced in 1925 and named Texas) had yet been built with anything larger than 63" drivers.

Larger wheels were needed for two reasons. First, small drivers on big engines lacked sufficient space for enough counterweight to properly balance these wheels. This was a big issue with 2-10-2s and early 2-10-4s. The second reason related to horsepower. A superheated engine could normally be expected to develop its peak power at piston speeds of about 1,000 feet per minute. In fast freight service, a locomotive needed to be in the high-end range of its horsepower curve between 40 and 60 mph. Piston speed is a function of both stroke and driver diameter. So, with the common piston strokes in use (30" to 34"), a driver diameter of 69" or greater was needed to meet this requirement.

The first Berkshire with large drivers was built for the Erie Railroad in 1927; it had 70" wheels. Nickel Plate Road's famous 2-8-4s had 69" drivers. Early 2-10-4s were drag engines — while they had powerful boilers, their running gear was not a big improvement over the 2-10-2. The Chesapeake & Ohio T-1 of 1930 changed this, coming with 69" drivers. The 2-10-4s with the biggest wheels (74") were Santa Fe's 5001- and 5011-class engines.

Articulateds designed for fast freight service also had large drivers. All 4-6-6-4 Challengers had either 69" or 70" drivers; the Big Boy had 68" wheels; Norfolk & Western's A-class 2-6-6-4 rolled on 70" drivers. But, thanks to the enormous girth of its boiler, the 2-6-6-6 Allegheny had to get by with just 67" drivers.

Timken's "Four Aces" 4-8-4 of 1930, built by Alco, was the first steam locomotive with roller bearings. It's shown here on Pennsylvania's Harrisburg-Pittsburgh main line. *Timken*

Combustion chambers

At the beginning of the Super-Power era, the value of a combustion chamber was not fully appreciated. Combustion chambers were not new — they dated from before World War I—but early ones were of riveted construction and had a reputation for leaking and high maintenance. This problem was eliminated with the universal acceptance of welded firebox construction in the 1920s.

Locomotive builders had long understood the need for plenty of firebox volume, but there was little agreement about how much was enough. The combustion chamber — a space in the boiler between the firebox proper and the rear tube sheet — added volume without contributing additional grate area. It promoted combustion by providing more space for the mixing of fuel and air, an important consideration as combustion rates increased. Nonetheless, not until the 1930s were locomotives built with truly big combustion chambers for maximum performance.

In boiler design, a trade-off existed between tube length and combustion chamber length. Adding to the combustion chamber improved combustion, but shortening the tubes to accommodate it reduced the amount of available heating surface. However, the heat absorbed per unit of tube length is exponential — i.e., the tube segments closest to the firebox absorbed the greatest amount of heat, and the tube segments nearest the smokebox absorbed very little. So, if tube lengths were shortened to accommodate a combustion chamber, the portion removed was effectively the segment that absorbed the least heat. Consequently, with coal-burning locomotives it was almost always beneficial to sacrifice some tube length for more combustion chamber. As a rule, tube lengths for non-articulated engines need not be greater than 19-20 feet.

With oil-burners, it was a little different. An oil-burner's firebox does not absorb as much heat as a coal-burner since an oil fire is less radiant. Consequently, fire gases entering the tubes are normally hotter than for a coal-burner, and it was usually beneficial to give an oil-burner an extra couple of feet of tube length to absorb this additional heat. Tube lengths of 21-22 feet were common.

The Lima A-1 had a tube length of 20 feet with no combustion chamber. Erie Berkshires had tube lengths of 21 feet, also without combustion chambers. Both had large grate areas (100 square feet). Nickel Plate Berkshires, with 90-square-foot grates, had tube lengths of 19 feet and 42" combustion chambers. All Texas types had combustion chambers. With their big boilers, there was plenty of room for both combustion chambers and long tube lengths.

All 4-8-4 Northerns were built with combustion chambers. Like the Texas types, their boilers were big enough for both a combustion chamber and long tube lengths. Likewise, all Super-Power articulateds had combustion chambers. Articulateds had a weight-distribution problem, so it was not uncommon to see tube lengths as long as 24 feet in order to move weight forward onto the front engine. The Big Boy has a 112" combustion chamber.

Turning up the pressure

There had been a trend toward higher boiler pressures since about 1900, and by 1910, pressures of 200 psi were commonplace. By the mid-1920s, many new locomotives had pressures of 250 psi, followed in 1930 by a Santa Fe 2-10-4 at 300 psi. In 1937, Santa Fe and Kansas City Southern took delivery of 2-10-4s with pressures of 310 psi, which would be the limit in North America for locomotives with radial-stayed fireboxes (the Big Boy uses 300 psi).

High-pressure steam had many advantages compared to steam at lower pressures. It was denser, more fluid, and contained more heat energy. A high-pressure boiler was more powerful. For example, a 300-psi boiler is about 5.5 percent more powerful than the same boiler operating at just 200 psi. Another advantage came from the use of smaller cylinders. Given two locomotives of the same tractive effort, one with a pressure of 250 psi and the other with 300 psi, if the first engine had a cylinder diameter of 27", the second one would need a diameter of only 25". Smaller cylinders permitted a locomotive to use less steam to do the same work. A smaller piston also meant less reciprocating weight that needed counterbalancing.

While high boiler pressures had advantages, there were also drawbacks. Boilers had to be built stronger. In the area around the firebox, the number of staybolts had to be increased. The boiler sections had to be either rolled from thicker steel, if ordinary carbon steel was used, or from higher-strength alloys such as nickel steel. Thicker carbon steel substantially increased locomotive weight; alloy steels were subject to caustic embrittlement. Boiler water impurities caused some alloy steels to become brittle and crack at stress points around the rivets. While this was minimized through the chemical treatment of boiler water, it was never solved completely until the Interstate Commerce Commission permitted all-welded boilers (which came too late to affect locomotives built in the 1940s). Many locomotives with alloy steel boilers were retired early because of stress cracking.

Roller bearings

In 1930, Alco built a 4-8-4 for the Timken Roller Bearing Company. The locomotive, Timken No. 1111 (the “Four Aces”), was a demonstrator equipped with roller bearings on all locomotive and tender axles, and its frame was integrally cast with the cylinders. The use of roller bearings required greater alignment precision, which the integrally cast frame provided. Roller bearings completely contained the driving axle and could absorb thrusts both vertically and laterally. They reduced driver pounding and ran cooler than the standard friction driving box. Beyond the need for lubrication and inspection, they could go more than 100,000 miles without attention. Roller bearings were one of the reasons for the outstanding availability and reduced shopping of Super-Power locomotives equipped with them, including the Big Boys.

At first, roller bearings were only used on passenger locomotives, as they were quite costly. Some railroads never used them, but other railroads did understand their value. N&W’s first Y-6 and A-class engines of 1936 came with roller bearings. They quickly proved their worth, and all new road power N&W acquired thereafter was so equipped.
Roller-bearing driving boxes soon became de rigueur for most railroads.

Roller-bearing installation did not stop with axle bearings. Several railroads had locomotives with lightweight roller-bearing-equipped main and side rods. Both the Pennsylvania and UP experimented with roller-bearing rods in the 1930s, but the first significant application came in 1937 with five streamlined NYC J-3 Hudsons. The PRR and N&W followed in 1942 with their respective 4-4-4-4s and 4-8-4s. NYC’s Niagaras of 1946 were so equipped, as were several C&O engines after the war. Nonetheless, the application of roller-bearing rods never extended beyond a handful of locomotives. They were largely viewed as an extravagance, and by the time they could prove their worth, the diesel was on the horizon.

New York Central's 4-8-4 Niagaras were equipped with roller-bearing side rods when built in 1946.
New York Central

Baker vs. Walschaerts

In the 1920s, long-travel valve gears were in vogue. Long travel (more than 8"), combined with wide steam lap (more than 1½"), improved valve events by keeping steam ports open wider and longer during the admission and exhaust periods. This improved the ability of steam to flow in and out of the cylinders unimpeded, which in turn reduced pressure losses. This was especially important for locomotives operating at speed, since the valve ports were uncovered for just a fraction of a second. Another advantage with long travel was that the valve spool was moving very rapidly at the time the steam ports were opened or closed, so the valve events occurred crisply. (This is one of the things that gives a locomotive a sharp-sounding exhaust.)

Cutoff, steam lap, and valve travel were interrelated. If steam lap was lengthened, without sufficient valve travel, cutoff would be limited. This was the situation with the A-1. Its steam lap was a very wide 2⁷⁄₁₆". Even though valve travel was a long 8¾", with this steam lap, the resulting maximum cutoff was just 60 percent. The A-1 was designed in this way to have both good valve events and to obtain the advantages of limited cutoff for slow-speed operation.

During the late 1920s and early '30s, Baker gear became popular, particularly on 4-8-4s. Baker gear had the advantage of comfortably providing longer valve travel (8½") than was available from Walschaerts (7½"). Nonetheless, Walschaerts ultimately remained the more popular of the two. It was a simple, straightforward design, and long valve travel gear did have some drawbacks. For one thing, it meant high valve speeds, so it took more work to overcome the frictional and inertial effects of moving the valve. As valve speed increased, the work required increased exponentially. A large locomotive at speed could easily consume more than 100 h.p. just to operate the valve gear. Because of this, early Baker gear suffered maintenance problems owing to pin and bushing wear.

When boiler pressures were down around 200 psi, valve diameters had to be large enough to permit the flow of the less-dense steam through the ports without undue restriction. Diameters of 14" were very common, and on some large 10-coupled engines with big cylinders, valve diameter could grow to 16". When boiler pressures were increased and cylinder diameters could be made smaller, it was also possible to reduce the valve diameters. Twelve-inch valves were used on many modern engines.

In the 1930s, two schools of thought developed regarding valve design. One believed that through use of high boiler pressures, valve diameter and travel could be reduced with no serious effects on performance. Most engines designed with this philosophy in mind had 12"-diameter valves driven by Walschaerts gear with 7½" travel. The smaller and shorter travel valves took less power to operate and placed less stress on the valve gear. The majority of railroads fell into this category; the Big Boy has Walschaerts valve gear with 12" piston valves.

The other school took an opposing view, emphasizing performance. These roads stayed with 14"-diameter valves with long travel and wide steam lap driven by an improved version of Baker valve gear. While fewer railroads adopted this approach, they owned some impressive locomotives. The NYC Niagara and the N&W J were in this category.

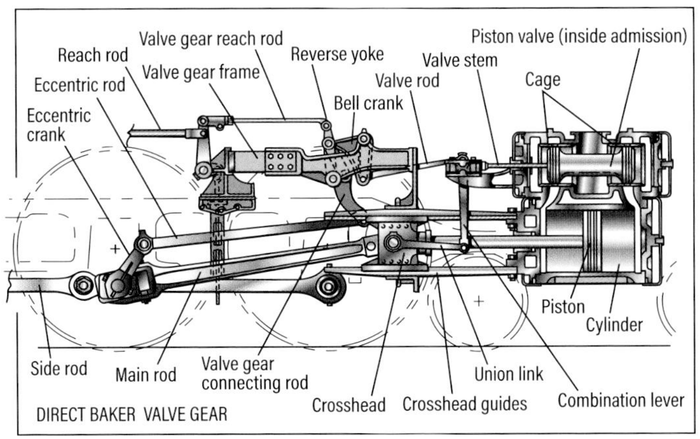

DIRECT BAKER VALVE GEAR

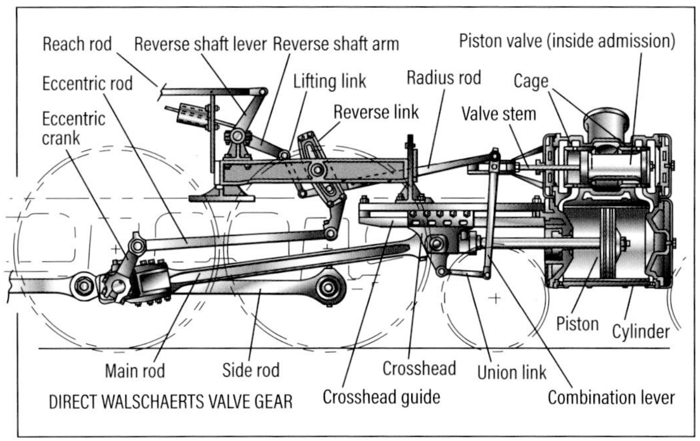

DIRECT WALSCHAERTS VALVE GEAR

The cross-balancing act

As locomotives grew larger and more powerful, the precision with which the driving wheels had to be balanced also increased. On early, small locomotives, just the rotating weights were balanced. As they got bigger, balancing the reciprocating weights became a requirement. Finally, a motion-induced nosing problem was minimized through a technique called cross-balancing.

Balancing the weight of the revolving parts is what is normally thought of when referring to counterbalancing. The revolving parts consist of the side rod, the crank pins, the eccentric crank, and the revolving weight component of the main rod. This was balanced by adding counterweight near the rim of each driver on the side opposite the crank pin. The revolving weights could be very closely balanced for any speed.

The second balancing requirement was to counter the effects of the reciprocating weights of the piston, piston rod, crosshead, and the reciprocating component of the main rod. As these masses moved back and forth, they caused a jigging motion in the locomotive that became very pronounced. To counter it, additional weight was added at the same locations on the driving wheels as the weight previously added to balance the revolving parts.

The additional weight, however, caused a new problem, since it overbalanced the wheels with respect to the rotating weights. The resulting rotational imbalance tended to hammer the rail at the bottom of its revolution while at the top it would tend to lift the wheels and potentially cause high-speed slipping. Its severity increased with speed. It got the name "dynamic augment" since these so-called hammer blows augmented the static axle load (the weight at rest) of the locomotive. Dynamic augment could damage rail, and it created a rough-riding locomotive.

Cross-balancing, or dynamic balancing, dealt with the tendency of a locomotive to begin side-to-side nosing as speed increased. The source of the problem was that the driving-wheel counterbalance weight was located a good deal inboard from the revolving weights (rods, etc.). As these two weights rotated on each side, they imparted forces of a similar magnitude, but at different points on the axle because they were separated by anywhere from 8 to 10 inches. Picture an axle in space with these separated forces and you can visualize that the forces would cause it to spin endwise. Now, if you place the axle back on the locomotive, you can understand that its rotation causes a twisting action on the locomotive frame. This induces a yawing force and causes the engine to nose from side to side. The problem was solved by adding counterweight to the opposite wheel, hence the term cross-balancing.

Cross-balancing became popular in Europe in the 1920s, but was not adopted on this side of the Atlantic until the 1930s. In North America, generally just the main driver was cross-balanced. Its implementation was easily seen by noting whether the counterbalance on the main driver was cocked a bit in the direction of the crankpin on the wheel on the opposite side. On locomotives with Boxpok drivers, a separate weight was used. Many older locomotives were rebuilt with cross-balancing.

End of the era

The innovation period of Super-Power steam was rather short. It began with Lima's A-1 in 1925 and, it may be argued, ended with the introduction of the C&O Allegheny and UP's Big Boy in 1941 — just 16 years later. These locomotives were the last to embody new features such as new wheel arrangements and massive boilers. Additional Super-Power locomotives were built after 1941, but they employed existing technology.

Steam technology received a big boost in the 1920s with the creation of several new wheel arrangements and the advent of many new improvements and appliances, but the Great Depression ended most of this development. Concurrent with the revival of the nation's economy in the late 1930s, locomotive innovation also began again. Several new designs were conceived including high-speed 4-8-4s, Challengers, the Big Boy, and the N&W A. The U.S. entry into World War II terminated new development, and after the war it was clear to most that the diesel would ultimately reign supreme. While the PRR continued to tinker with its unsuccessful duplex designs — and the N&W professed faith in steam — most other railroads rushed to dieselize. By early 1960, the last Super-Power steam in regular service had been set aside, and the era had ended.

Neil L. Carlson is an electrical engineer with a life-long interest in steam locomotive technology. This chapter was adapted from "Super-Power," a two-part article that appeared in the May and June 2000 issues of TRAINS.

Union Pacific's Big Boys utilized many Super-Power concepts. The coming of the diesel locomotive limited further advances in steam technology. *Trains magazine collection*

7 Yes, They Could Handle It

by Jim Wrinn

How UP acquired No. 4014 from a California park, towed it 1,300 miles, and rebuilt it

FOR YEARS, CONVENTIONAL WISDOM was that a Union Pacific Big Boy locomotive either could not, should not, or would not be restored to steam. It was too big. Too expensive. No need for it. There was no place to turn it. It would crush the very roadbed upon which it trod. Its consumption of fuel and water would be animalistic, at best. The list was long and ranged from the probable to the absurd.

But then a strange thing happened. Someone in a position of responsibility took a fresh look at the idea. Instead of accepting conventional wisdom — from know-it-all fans to grizzled railroaders — that an operating Big Boy would be a bad thing, that person gave the prospect an opportunity. And what that person recognized and decided has supercharged UP's steam locomotive public relations with the return of Big Boy No. 4014.

More on that later, but first, we should start with the most basic question of all: How did this happen?

The answer is that about 2011, a West Coast businessman, with a knowledge of and a background in railroading, approached the UP about restoring one of eight surviving Big Boy locomotives. He even offered to pay for the restoration that would obviously run into the millions. Given who it came from, it wasn't a bluff, and the idea caught the interest of executives at UP's headquarters in Omaha, Neb. They put the question to Ed Dickens, who had just taken over command of the company's steam locomotives — its Heritage Fleet, as UP calls its steam public-relations effort.

"I was approached as to whether this was feasible, and from that conversation, we began our search for a candidate," says Dickens, senior manager of Heritage Operations. He was asked if the steam crew was interested in adding such an engine and if it would be feasible to operate the locomotive in the same service that UP operates its two other steam locomotives. "Knowing the infrastructure of the UP, I felt this was something that could be done," he added in a 2013 interview. "The opportunity is right."

The more he looked into the idea, the more he liked it. Having just changed managers, it would give the program a new and exciting direction. It also would take the company's use of its steam bravado to new heights.

The railroad had run 4-8-4 No. 844 since 1962 and 4-6-6-4 No. 3985 since 1981 as public relations and goodwill ambassadors for its far-flung system across the Western U.S. But the prospect of a legendary Big Boy as part of the program was now a possibility. With this much lead time, there was also the possibility of completing the restoration in time for the 150th anniversary of the first transcontinental railroad in May 2019. Now the questions began to shift. They were not if the railroad should do this but how. Of the eight survivors, scattered from Pennsylvania to Wisconsin and to Texas (see page 196), which one should the railroad go after? Which was mechanically in the best condition? Which politically and

Stripped down to the bones! Big Boy No. 4014 rests on steel I-beam cribbing and a 100-ton freight car truck inside the cavernous back shop in Cheyenne, Wyo., in March 2017. With all jacketing and appliances removed and the smokebox door open, it's time to get inside and work on the massive 300-psi boiler. Inset: The sign above the steam shop offices proclaims the steam crew's pride in the restoration of a Big Boy locomotive in the heart of historic Big Boy country. *Jim Wrinn*

UNION PACIFIC 4014
UP 4014
PROJECT
CHEYENNE
STAR 22
Haulotte

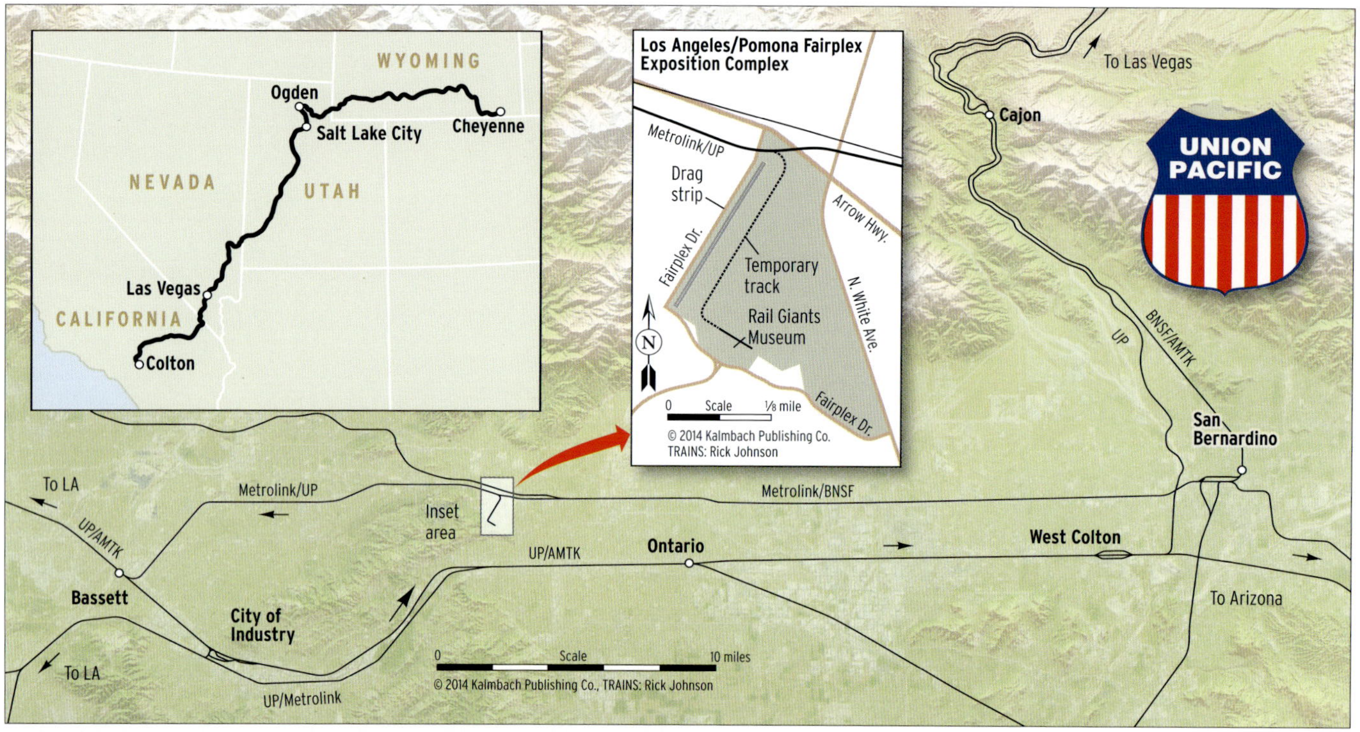

Spray-painted plywood signs speak of a sense of humor as well as a recognition of the challenges the UP steam team faced in 2013 when extracting Big Boy No. 4014 from its California home since 1962. *Jim Wrinn*

practically would be available? Would any organization possessing one of these last giants be willing to relinquish it?

The search quickly led to No. 4014, which had been owned by the Southern California Chapter of the Railway & Locomotive Historical Society since 1961 and on display at the Los Angeles County Fairgrounds in Pomona since 1962. Its home, appropriately, was called the Rail Giants Museum. It was the star of the show in an exhibit that included UP 4-12-2 No. 9000, a Santa Fe 4-6-4, and many other smaller locomotives assembled over decades. Two volunteers, Paul Guercio and Rick Brown, had been especially taken in by the Big Boy, and they did more to keep it lubricated, cleaned, and ready, as if one day, someone might come looking for it. Their actions were nothing short of prophetic.

Dickens says he looked at other survivors but No. 4014 immediately became the front-runner because its boiler barrel — the beating heart of a steam locomotive — was in excellent condition. The relatively dry desert climate of Southern California had turned out to be an excellent place to lay up a locomotive. "There's some pitting, but overall it is in extremely good shape," he said of No. 4014 in 2013. "The interior of the boiler is like a time capsule. I could see writing on the steel and as I worked my way inside it during an

November 13, 2013, will always be known as the day when No. 4014 started down the road to restoration. A front-end loader tied to the tender coupler gently tugs the dead locomotive backward from its display site at the Los Angeles County Fairgrounds and the Rail Giants Museum. *Three photos, Jim Wrinn*

The steam crew checks the position of the No. 1 driver wheel on temporary track as it inches out of its museum berth. Plywood under the ties was installed to keep level on undulating parking lot surface.

inspection, I could see a fuseable plug that looked like it had been removed yesterday, but it hadn't been."

The 300-psi pressure vessel that constitutes the Big Boy boiler is critical, and its display in a relatively dry climate meant the locomotive had suffered little damage from rust and condensation over the years. "On a lot of museum or display locomotives, you see moss or grass growing out of the jacket," he said. "That's not the case here." The engine was also largely intact, down to nozzles for the injectors. It had not been robbed of parts to keep other engines running or in cosmetic condition. Much of the air-brake system was complete, a good thing considering the engine had to travel dead-in-tow more than 1,200 miles from Southern California to the UP steam shop in Cheyenne, Wyo.

Some members of the chapter's board of directors agonized over losing their biggest locomotive. On the UP, Dickens kept his team excited about their potential new addition to the steam fleet. As they would ride in No. 844's cab, where the roar made conversation impossible, Dickens would hand gesture four, zero, one, four using his fingers to remind them of what was ahead.

After weeks of negotiations, UP in 2013 struck a deal with the historical society chapter

Building track in the parking lot of the Los Angeles County Fairgrounds, the UP steam crew fashions a curve for Big Boy No. 4014 to traverse as it makes its way out of the Rail Giants Museum in Pomona, Calif.

to trade the Big Boy for an operating 1970s diesel locomotive — SD40-2 No. 3105 — a steel caboose, and proceeds from a future excursion with the restored No. 4014 to be run out of the Los Angeles area.

First inches, then feet, then miles

That bargain solved one problem and created another. A big one, in fact. UP's new prize member of its steam stable was trapped behind several other locomotives at the Rail Giants museum. It was almost a mile across a parking lot at the fairgrounds to the nearest railhead on a Metrolink commuter line. In short, it was marooned at the farthest point away from active rails.

To extract the engine, UP took on a herculean effort. The steam crew mounted a compressor on the tender and reactivated the air-brake system to give the engine stopping power. The crew installed panel track across the parking lot and used a giant earthmover to drag No. 4014 ever so slowly backward over a period of weeks to a point where crews could stage the engine for a daring nighttime dash on Los Angeles' Metrolink commuter railroad. The effort began in November 2013, and the midnight dash took place the following January.

Over the next few months, UP prepared to tow its repatriated Big Boy to Cheyenne. That journey took place over a two-week period in April and May 2014. A pair of handpicked, modern freight diesels, SD70ACes Nos. 4014 and 4884 (like Big Boy's wheel arrangement) moved the engine from Los Angeles to Ogden, then across original Big Boy territory to Cheyenne. Everywhere the locomotive and its diesel accompaniment went, people turned out. In Salt Lake City and Ogden, thousands came out, some from hundreds of miles away, just to see the dead engine. The engine returned to the places where it had become famous: Wasatch Grade, Peru Hill, Sherman Hill. The journey concluded on a rainy afternoon in Cheyenne when diesel No. 4014 tucked steam No. 4014 inside the UP steam shop next to Living Legend No. 844.

With step one, acquisition of a Big Boy locomotive, complete, it was time to turn to the hard work of restoration. Progress, however, didn't begin right away. No. 844 was lame and had to have a full boiler rebuild. While that went on, Big Boy sat. Two years went by with little direct work, but significant efforts to upgrade the shop and install an overhead crane. When the crew made parts for No. 844, they also made a set for No. 4014. When they ordered equipment for No. 844, they ordered for No. 4014 as well. Immediately after No.

Near Etna, Nev., just south of Caliente, the No. 4014 deadhead train rolls northbound after leaving Las Vegas. Behind the second diesel are tool cars, crew cars, and flatcars for additional braking power. Next stop, Milford, Utah, on the famous route to Salt Lake City. *Jim Wrinn*

123

Moapa, Nev., some 50 miles northeast of Las Vegas, is a strange place for people to show up next to the Union Pacific main line. But in spring 2014, they did just that to see dead No. 4014 on its way home to Cheyenne and new life. *Two photos, Jim Wrinn*

844's return to operation in July 2016, the steam crew wasted no time in getting to work on its next project. They used the 4-8-4 to take No. 4014 to the turntable at the steam shop. They gave the new addition a spin so that No. 4014 faced west. They separated the Big Boy from its tender and backed the engine into the cavernous steam backshop.

Finally, it was game on. The task ahead was monumental. Even though No. 4014 had been pronounced sound, it was still a locomotive that had more than a million miles on it, shuttling between Ogden and Cheyenne for almost 20 years with heavy tonnage trains on the punishing grades of Wasatch and Sherman Hill. It also had sat outside, exposed to the elements, for almost six decades.

The first thing in any major steam restoration is to disassemble the engine and begin a thorough boiler inspection. That means removing the tubes, flues, and superheaters, testing all of the bolts and rivets, checking the metallurgy, inspecting the seams of the boiler courses, and performing an ultrasonic test to determine the thicknesses of the boiler shell and the firebox walls. With plans to convert No. 4014 to burn oil, the crew removed the

Union Pacific diesel No. 4014 backs steam locomotive No. 4014 into the steam shop at Cheyenne on May 8, 2014. For the first time in 55 years, 4-8-8-4 Big Boy 4014 had come home, preparing to begin its next career.

ashpan and coal-burning grates that would no longer be needed. Some grates were returned to the Rail Giants museum for display, and others were donated to the Milwaukee Road 4-8-4 No. 261 in Minneapolis for use in another Alco-built steam locomotive still running.

December 2016 brought visible and significant progress. In that month the steam crew contracted with a derailment-cleanup company to bring two heavy-duty cranes to the shop and hoist the 95-inch-diameter boiler into the air, enabling the front engine and the forward part of the articulated frame to be separated from the rest of the locomotive. The front portion of the boiler was set onto a specially constructed dolly.

This step was required, says Dickens, because the 4000-class design makes many areas inaccessible without taking the locomotive apart. An almost rivet-for-rivet disassembly is the only way to properly analyze and repair many components.

It had been the late 1950s since a Big Boy locomotive had been torn apart for rebuilding, and the heritage team marked every step of the way. "It might sound corny, but we've

X4014
4014
UNION PACIFIC
4014
PLEASE KEEP OFF

had a lot of firsts," Dickens says. "Taking the 4014 apart, being the first to walk underneath the boiler — some of these things haven't been done in 60 years. You can go and look at a Big Boy in a park, but that's very different than actually disassembling one of them."

The bulk of the work would be done in house by the UP steam crew. Few components were sent to contractors. The eight driving-wheel sets, which weigh 16,000 pounds each, were sent to the Strasburg Rail Road backshop in Pennsylvania for crank pin and axle work and installation of new tires. The two cross-compound air pumps went to Colorado's Back Shop Enterprises, run by Bernie Watts, for rebuilding.

Cheyenne, Wyo., is known for several things. It's the state capital. It's also just east of the Rockies and in the spring it is a wind-whipped place. That's what we found in March 2017 when TRAINS magazine staff visited the steam shop. Outside, gusts of up to 60 mph made it difficult to stand; inside, the steam crew was quietly assessing No. 4014's condition and machining replacement parts.

With the locomotive partially disassembled, the crew made progress on repairing pistons and cylinders, running gear, wheels, and other parts. Returning these components to pristine condition requires "thousands of man-hours," in Dickens' words, but overall he was surprised by how little work the locomotive will require.

As of March 2018, the time of our next visit to the shop, No. 4014 was about as disassembled as you can get and still be counted as a locomotive. The wheels were out from underneath it — all 24 of them — and 100-ton freight car trucks support the frame. It held the title of the world's largest 0-0-0-0. The boiler was as empty as it has been since Alco built the 4-8-8-4 in 1941. The cab was in another part of the shop and resting near the 844. The tender, as yet un-rebuilt, was sitting in the yard. But there are signs of progress everywhere at the new Alamo of Steam.

The Alamo of Steam was a term that then-TRAINS editor David P. Morgan coined to describe the Roanoke, Va., headquarters of the steam-committed Norfolk & Western in the 1950s, and one that I appropriated and modified with the word "new" in 1995 to describe the UP steam shop. I wrote this in March 2017: "They are the kind of signs that give confidence to the crew that is laboring so hard to see this done, the railroad that wants to see this behemoth completed as a goodwill ambassador, and the fans who have long dreamed of a Big Boy back in steam.

"The crew is working hard on multiple tasks to get the job done. They're sequestered in a 'clean room' machine shop, where computerized machine tools of today are bringing back the past. They're huddled around the gigantic trailing truck frame (estimated weight, 17,000 pounds, and the size of a small car) that's been removed. They're perched in a lift on the side of the boiler. Respectively, they're making bolts, inspecting the casting, and welding. They're also deep inside the firebox, where the task of renewing metal is ongoing. It is relentless, exhausting, and exhilarating work all at one time. But you can sense that this is a crew that is proud of getting the 4-8-4 back on the road in 2016 and that is on the verge of something big, something that was always believed to be too big to ever happen, too far out of reach, or just a silly dream. But dream no more. Throughout the shop, confidence is high that we'll see No. 4014 done in time to celebrate the 150th anniversary of the Transcontinental Railroad in May 2019. That will be 60 years since a Big Boy last pulled revenue freight in regular service. It will be a cause of celebration among those who love steam the likes of which we've never seen before."

During that visit, Dickens said the work was about 50 percent done. The focus was on the 300-psi boiler that is the heart of this 7,000-hp giant of the rails. He estimated that 85 to 90 percent of the parts needed to restore the locomotive are on hand. Boxes of new staybolts of varying lengths were awaiting installation.

I reported further: "The steam exhaust water pump sets ready on a workbench. A Nathan lubricator shines in fresh gloss black, having been overhauled. The much celebrated burner for oil firing is in the building. The crossheads are nearly done with machining. The brake stand has been rebuilt. The list is a long and tedious one, but such is the way of steam locomotive restoration for the main line on a stage that only Union Pacific could provide. Only one highly visible part of the old No. 4014 — the ashpans from the locomotive's coal burning days — will go away with the conversion to oil firing. The time for taking parts off Big Boy has passed, and the time for putting new or refurbished parts back on has arrived.

"That is extremely evident one track over from the Big Boy, where the front engine awaits wheels and a boiler: The front pilot has been renewed. Rebuilt cross-compound air pumps are installed on the platform. Lubrication lines, soon to be covered up by the boiler, snake across the top of the frame in all directions.

Union Pacific steam manager Ed Dickens, right, finishes up paperwork in the moments after No. 4014 was parked inside the steam shop for the first time. At right is No. 844, in the middle of boiler work. *Jim Wrinn*

With the cab removed and the backhead exposed, work progresses on No. 4014's boiler in March 2017. Compare this image to the photos on page 149 showing the cab being installed in this area. *Four photos, Jim Wrinn*

Looking down the barrel of the long boiler from the smokebox end, and with both front and rear tube sheets removed, boilermaker Jimmy Thompson uses a cutting torch to remove a staybolt.

We're far from Cheyenne, Wyo., in the heart of Pennsylvania Amish Country, in Strasburg, Pa., where the famed Strasburg Rail Road shop was working on No. 4014 wheel centers and axles in May 2017.

"Soon, the drivers with new tires and crank pins will return from rebuilding at the Strasburg Rail Road. Workers will load the tubes and flues into the boiler. Reassembly will begin with thousands of parts ready to breathe new life into this old friend."

In June 2018, we visited once more. My report that time focused on the work that had been accomplished in three months and the work that was ahead: "With six of the eight wheel sets back from renovation and application of new tires, boiler work at the company's steam shop in Cheyenne, Wyo., has continued at a rapid pace. Rigid staybolts, flexible stays, rigid crown bolts. Rivets. Appliance studs. The gritty, tedious work continues on the 300-psi pressure vessel. In late June, the nine-member steam crew was installing the last rigid staybolts in the firebox. The rear truck that supports the firebox and cab was ready, and attention turned to the cab, which is getting significant amounts of new metal. Ahead is rewheeling of the locomotive, installation of tubes and flues, and a hydrostatic test of the boiler to ensure its integrity."

Soon the crew was boring the cylinders and replacing the bottom of the cab where rust had taken its toll. By December 2018, No. 4014's restoration was moving forward at a rapid clip. The engine was moved outside to be reunited with its rebuilt front and rear engines.

In a scene unimaginable only a few years prior, No. 4014's boiler and front engine share space inside the Cheyenne backshop. New lubrication lines (light colored strands) have already been run on the engine.

The steam shop crew pauses during work on No. 4014. From left, Troy Plagge, Ed Dickens, Austin Barker, Kirt Clark, Jimmy Thompson, Bruce Kirk, Ted Schulte, and Garland Baker pose with No. 4014's 68-inch Boxpok drivers in June 2018. *Two photos, Jim Wrinn*

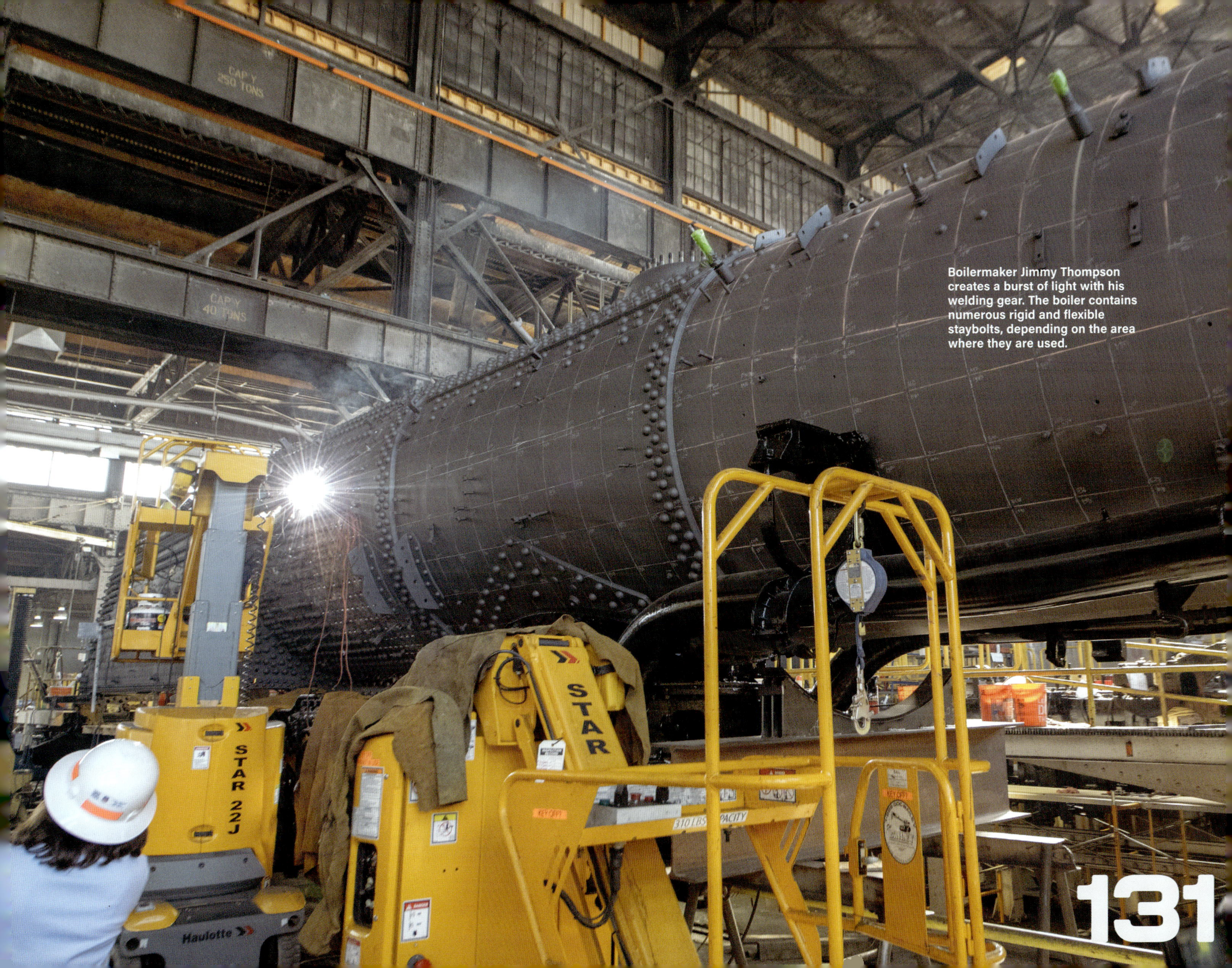

Boilermaker Jimmy Thompson creates a burst of light with his welding gear. The boiler contains numerous rigid and flexible staybolts, depending on the area where they are used.

Looking from inside the firebox at work literally in the belly of the beast. Our perspective here is looking back toward the backhead. The opening is the firedoor. On the floor is ultrasonic testing gear. *Three photos, Jim Wrinn*

The almost-completed boiler was only missing the steam dome cover and smokebox inspection hatches. The backhead neared completion with valves, piping, water glasses, and other controls. But the real drama was down at track level. Side-boom cranes lifted the boiler while the front engine was pushed into place. A wedge was clamped onto the track to keep the No. 4 driver in place while the boiler was lifted at an angle that allowed the articulation tongue and hinge to go back together. At the same time, the crew managed to get the steam exhaust pipes and the front boiler wear plate into position. It was a coordination of all of these aspects that got the Big Boy back into one piece and looking like a locomotive once more.

Meanwhile, a notice published in the federal register in December 2018 showed that UP had asked the Federal Railroad Administration for waivers on Positive Train Control compliance for steam locomotives Nos. 844, 4014, and Challenger 4-6-6-4 No. 3985,

Big Boy parts on a workbench: At left, mechanical lubricator; at right, a brake cylinder housing. Both were being rebuilt to prepare No. 4014 for a long service life.

We are looking across three sets of drivers in June 2018 that have just returned from quartering and installation of new tires at Pennsylvania's Strasburg Rail Road. In rear is the boiler backhead.

The rear trailing truck, sans wheels and resting perpendicular, fills the foreground of this view of No. 4014's massive boiler in the steam shop in March 2018. Boilermaker Jimmy Thompson rides a lift on the side.
Jim Wrinn

The Big Boy firebox, from the engineer's side looking forward toward the smokebox, shows multiple circulator tubes coming off the crown and sidesheets, staybolts, combustion chamber, and rear tube sheet. *Two photos: Jim Wrinn*

We are at the smokebox end of No. 4014 in March 2017 looking back through the boiler. Front and rear tube sheets create a pattern while a worker removes a defective staybolt in the firebox.

It's March 2018, and much of Big Boy No. 4014's front engine frame has been completed. Lubrication lines, inaccessible once the boiler is reattached, are wrapped in light-brown insulating covers. The boiler sits one track over to allow workers to make progress on both portions at the same time. The grid on the boiler is made in soapstone so that each section can be tested for integrity. Mechanical lubricators have green tops. The perspective is from near the second set of cylinders looking forward. *Jim Wrinn*

which is stored out of service in Cheyenne. Says the filing: "Union Pacific states these units are specifically designated and utilized on an extremely limited number of excursion trips annually. Excursions utilizing this equipment historically and continually involve the highest level of preparation in planning and operational safety during operation, giving special designation to each move. No less than two operators are present at any given time ensuring continual functional safety awareness during operational moves. UP explains they have been operating steam-powered excursions without PTC-initiated application and have found no adverse mechanical effect on operational safety." The FRA decided that the locomotives do not need a waiver as UP is still on track to implement PTC by the end of 2020.

Big Boy work continued through the winter with the installation of tubes, flues, and superheaters — the guts of the impressive boiler. Cut to precise lengths, the tubes and flues are installed, rolled, and then welded into place. Set the superheaters — coils of piping in which steam runs to get a final boost before being released to the cylinders — in place and you've got yourself a boiler that is on the verge of completion. All it needs is a final squeeze test to make sure it is sound.

On Feb. 6, 2019, UP reported a breakthrough: That final squeeze — a successful hydrostatic test of the Big Boy boiler, the heart of the locomotive. A hydrostatic test is required by Federal Railroad Administration rules to confirm the integrity of a steam locomotive's boiler. It confirms that the boiler will be able to handle the heat and pressure of steam. Such a test requires the boiler to be filled with warm water and the pressure to be raised 25 percent above the maximum allowable working pressure. In Big Boy's case that is 375 psi. A typical rule of thumb in steam restorations is that it takes about six months of work after a hydrostatic test to complete all of the other work necessary to start operating. That meant UP had a tight window in which to complete the locomotive before the May 10 celebration of the Golden Spike's 150th anniversary. While the hydrostatic test took place, other crew members were busy running air and steam lines and piping up the No. 8 ET brake system in the cab. One of the good things about Big Boy's immense size is that you can spread out a restoration crew so they don't get into each other's way.

A few weeks later, UP announced steam testing had begun inside the shop. On April 9, the railroad said that the restoration had reached a significant milestone: The engine had been fired up for the first time in 60 years. Given that No. 4014 had not burned oil before, the crew took its time to adjust the burner so that it evenly heats the firebox sheets.

Staybolts by the box are placed in containers for transfer from the machine shop to the shop floor. *Three photos, Jim Wrinn*

Garland Baker with the computer-controlled machine that cuts staybolt threads. The massive Big Boy may be a 1941 product that was state of the art for its time, but some of its parts were made with the most modern of machines

It was now coming down to the deadline UP had set for Big Boy. A May 4 trip to Ogden for the sesquicentennial of the Golden Spike had been set. There was plenty to do. Big Boy hadn't even moved under its own power yet. A planned trip by No. 844 to Ogden the week of April 27-28 was cancelled so the steam crew could focus on completing No. 4014's restoration.

The clock was ticking — and ticking louder by the minute.

Years of hard work paid off on April 27 when the crew hitched a tug to No. 4014 and pulled it out of the shop and into the sun. That afternoon, they worked steam on the locomotive, testing the safety valves, blowing out the valves and cylinders of dirt and debris, and ultimately testing the whistle. The pops blew in the windy Wyoming air. The sudden rush of steam flowing from the four cylinders and valves created a horizontal steam cloud.

The restored backhead of the locomotive shows firedoor, studs for appliances, and numerous staybolts. The backhead is jacketed and insulated before controls, water glasses, and other items are installed.

And a little after 5 p.m., General Foreman Austin Barker and Rail Giants museum's Paul Guercio sounded the giant's throaty voice for the first time in six decades. The Hancock three-chime, long-bell whistle announced to everyone standing on a nearby bridge and to all within earshot of downtown Cheyenne that Big Boy was back.

With the countdown on, there was no time to waste. Valves, pistons, and valve gear all had to be installed, and the valves precisely set for proper travel to ensure the accurate admission of steam to the pistons. A thousand and one time-consuming details go into the restoration of a steam locomotive, and Big Boy was no different than any other of its kind.

The following week was deadline week, and the crew spent the time inside the shop while an early spring snow and wind whirled outside. For days, there was no visible sign of activity to observers, just a few cars in the parking lot late at night.

Every moment counted at this point. "We were still bolting items on and loading the tool car," Dickens says of the last few hours of the work. "We knew we'd get it done.

The final reassembly included the lagging and jacketing, a tedious and painstaking job on any steam locomotive. True to form for a do-it-yourself team, the crew made the bands,

With a new crane overhead, No. 4014's boiler sits inside the cavernous Cheyenne, Wyo., backshop in June 2018. This place once housed Big Boy overhauls in the 1940s and 1950s and the steam shop for years. *Two photos, Jim Wrinn*

Steam shop worker Bruce Kirk checks a rigid staybolt on the engineer's side of the firebox. Such bolts allow the firebox to expand when hot. The other staybolt type is flexible and is often found on the roof.

Tenders: Swipe right or swipe left

Which tender would you pick to operate behind No. 4014: The one that had been behind Challenger 4-6-6-4 No. 3985, on left, or the one that came with No. 4014 from California, on the right? Both were photographed in the yard at Cheyenne in June 2018. *Two photos, Jim Wrinn*

When Big Boy 4014 first rolled under steam in 2019, it did not use the tender it brought with it from California. In the interest of time and expediency, the UP steam crew decided to prepare the tender tank from the Heritage Fleet's Challenger 4-6-6-4 No. 3985 to go behind the Big Boy for its inaugural season back in steam.

The reasoning is simple: No. 4014's tender is outfitted with a coal bunker and auger to feed its cavernous firebox. Number 3985's tender was converted to carry fuel oil when the locomotive's fuel was switched in the late 1980s. The two tenders are essentially the same with minor differences other than the type of fuel they carry. No. 3985 is safe inside the Cheyenne roundhouse, having been out of service since 2010. No. 4014's tender would have required the construction of a fuel oil tank and piping modifications. No. 3985's tank is also in good shape with paint and body work, while No. 4014's tender needs some sprucing up.

To prepare No. 3985's tender, the steam crew cleaned out hundreds of pounds of rust that had accumulated inside the water portion over the years. Centipede tenders are common to not only Challengers but also UP 4-8-4s like No. 844. They're of the 4-10-0 wheel arrangement with 42-inch-diameter wheels. They carry 28 tons of coal or 6,000 gallons of fuel oil, and 24,000 gallons of water (the last five Big Boys had tanks of 25,000 gallons of water). The centipede tender, more so than the long-legged locomotive, is one of the limiting factors of the engines on tight curves.

Interestingly, both tenders have sat on a siding just outside the steam shop at either end of a cut of cars.

Eventually the tanks will be returned to their respective locomotives. But for now, a bit of the much beloved Challenger, a popular member of the steam fleet since its introduction in 1981, will share the stage.

And if you're still concerned about the appropriateness of the swap, consider this: No. 4014 has carried the tender from sister Big Boy No. 4015 for years since the steam era. The tender that came with it from California carries number plate 25-C-116, which is the one from No. 4015.

When Big Boys were being shopped in the 1940s and 1950s, if a locomotive were ready to go and its tender was not, shop forces would grab a tender that was ready, quickly scratch out the old numerals and paint in the new one, and put the engine back on the road. They were needed to move freight, and nobody cared which fuel and water source accompanied these 7,000-hp power producers.

So now, you know what was behind No. 4014 when it emerged from the shops and what will be behind it in years to come: A tender from No. 3985 initially, and a tender from a close sister sometime in the future. — *Jim Wrinn*

The smokebox interior shows fabricated design, smokestack, and live steam tubes running from the superheater header. Plastic buckets sit on a covering where the exhaust nozzles are normally found.
Three photos, Jim Wrinn

UP steam shop manager Ed Dickens shows a rebuilt crosshead guide on a workbench. Note the many machines placed throughout the shop for easy access.

The biggest change in No. 4014 from its regular service career to restoration is the conversion from coal to oil firing. This is the burner for the oil firing apparatus. Fuel oil enters at right and is sprayed into the firebox as a mist that ignites in the air.

brackets, and clips for the sheet metal that would provide No. 4014 with its silky-smooth exterior.

Then they set the valves that time the admission of steam to the cylinders. The job is a tad easier given that Big Boy is a simple and not a compound locomotive. "Its basically two engines under one boiler," Dickens says. "That makes it slightly easier to set the valves."

Little has changed on the Big Boy from its regular service era. Gone are the stoker and the ashpan; both are unnecessary on a coal burner converted to oil firing. Another casualty of the change is the hydrostatic lubricator that was in the cab. It is gone — it only serviced the stoker motor, which, of course, has been removed. The superheater header to appliances is blanked off — they will perform just fine without a direct steam supply from the boiler's hottest steam. Otherwise, save for a few electronics such as cab signals and a radio, Big Boy is today about what it was in 1941.

The boiler, the heart of the locomotive is amazingly original. Some 800 of 4,300 staybolts were replaced, including many crown stays. A few firebox patches were installed — especially in the combustion chamber just forward of the firebox, where cinder erosion takes its toll. Both the front and back tube sheets were replaced. "When we started, I told management that I wanted to have confidence in the pressure vessel, and they supported that," Dickens says.

Timken went through all of the roller bearing assemblies on the locomotive. The shop crew installed 7,800 pounds of firebrick in the firebox. The most dazzling improvement may be in the cab floor, where a UP shield medallion 4014 is embedded in the waffle plate; it covers the drawbar pin and is removable for inspection purposes.

For its inaugural run, No. 4014's mechanical restoration took precedence over its cosmetic features. The guard over the chain drive to the lubricators and the replica builder's plates were still to be installed. Dickens says he still wants to nickel plate the throttle and the reverser. But for now, at least, the crew has produced a machine worthy of the main line. On deadline, they've completed the impossible. They've brought No. 4014 back to life.

How much did Big Boy's restoration cost? We don't know. The railroad will not say. But given the immensity of the project and the size of the locomotive it is not hard to estimate. Other locomotives the size of No. 844 that are under top-to-bottom restoration are coming in at in the range of about \$2 million to \$3 million. It would be easy to estimate that No. 4014 would come in at least twice that cost. For a locomotive that cost about \$263,000 when new, it is not out of the question to think that its restoration cost in the \$5 million to \$7 million range. To anyone who appreciates American history, the cost is worth it. And to the UP, the price is well within reach for the goodwill and favorable community relations that No. 4014 will earn from 2019 into the future.

With this massive overhaul completed, No. 4014 should be on the road as a UP goodwill ambassador for the next 45 years. With that, it is entirely possible that the engine will be in operation for its centennial in 2041. But we're about to get ahead of ourselves. Once the restoration work was done in April 2019, it was time to take Big Boy out for test runs, to break in a newly rebuilt engine. Such has been traditional on American railroads for as long as steam locomotives have been rebuilt, and UP has always followed that path as well. Those tests traditionally start in the yard, and move to the main line, where the locomotives can run, no gallop, like the horses to which they've been compared.

In No. 4014's case, she has been reborn to roam on the same route that she plied so many times between Ogden and Cheyenne. She has become special among 25 locomotives that were considered special from the day the first one was completed in 1941. She has become the last of its kind to run once more, a Big Boy back in steam!

This article appeared in Big Boy: Back in Steam, *a* Trains *special edition, in 2019.*

It's June 2018 and UP 4-8-4 No. 844, the only steam locomotive on a Class I American railroad that has never been retired, rests. At the right is No. 4014's cab, in the midst of getting new steel. *Jim Wrinn*

Cab installation: With boiler work completed, the boiler and frame are reattached and the crew cab is placed on the finished backhead. *Three photos, Union Pacific*

Center: A new shop crane eases the rebuilt and painted cab into place in February 2019. The blue flag indicates presence of shop workers on or under the engine. The circular pattern of rivets reinforces areas for air to be admitted to the firebox to aid in combustion.

Once more, Big Boy looks more like its old self. Note the rebuilt trailing truck — light-colored piping carries lubrication lines. The UP steam shop crew praises the new overhead crane installed to aid in the Big Boy and all future projects.

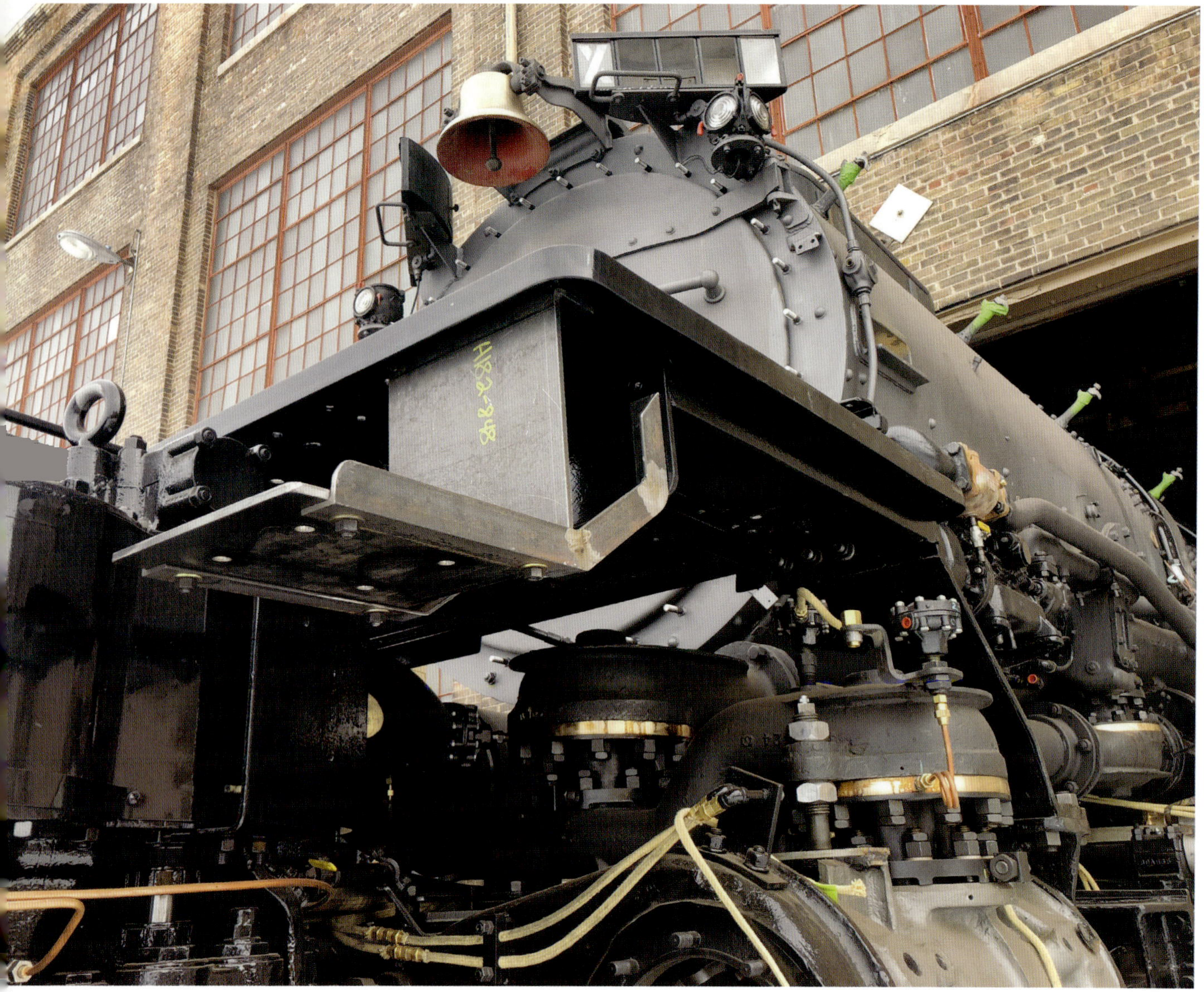

To reach the backhead to place the cab, the Big Boy was pushed out the west end of the Cheyenne backshop for the first time in February 2019. The completed locomotive begins to take shape. *Three photos, Union Pacific*

Center: It's a snowy February 2019 day, and we see Big Boy's face once more after three years inside the backshop. The restoration included work to renovate the pilot and retractable front coupler door, new air pump shields and radiator housing (the flat sheet metal on pilot), and rebuilt cross compound air pumps.

Still to come are the pistons and piston valves, fireman's side handrails, headlight, and UP shield emblem/front number plate that is one of the most visible portions of the engine. Overall, the Big Boy, even in the midst of completion, presents a pleasing face to all observers. The backshop once housed Big Boy work in the steam era.

8 On the road again

by Jim Wrinn

No. 4014 stretches its legs for the first time in 60 years

TEST TRIPS AND BREAK-IN RUNS ARE VITAL to the success of a machine that will be in the spotlight. Some chief mechanical officers want so many miles on their engine before they debut it for the public. Others want it to be in operation for so many hours. The end result is the same: a reliable locomotive.

In the case of Big Boy No. 4014, the railway preservation community watched and wondered when such a trip would occur as the calendar counted down to the May 4 inaugural run. With a giant machine to prepare, the work went down to the wire, and the crew buttoned up the engine, fired it up, and prepared it for a May 2 journey.

Originally, the test trip was to be a Cheyenne-to-Greeley, Colo., turn, on the main line to Denver. That would put about 120 miles on the engine. On the morning of May 2, the locomotive coupled up to a canteen, two tool cars, and a diesel on the trailing end to provide backup power and also to simulate a load. The train was prepared to go south, but owing to the steady pulse of traffic on the Overland Route main line and healthy business on the Denver line, the engine had a time getting out of the yard.

The assemblage finally left after 6 p.m., making the slow climb to the junction of Sherman Hill Track 3 and the Denver line in the fading twilight. The train stopped often so the crew could inspect, and it finally made it to Speer, Wyo., as darkness fell. The train got to Nunn, Colo., before stopping and returning. The crew knew the engine was good to go West.

The sun dips below the horizon as No. 4014 steams westbound out of Cheyenne, Wyo., for the first time in 60 years, on May 2, 2019. The location was the site of Tower A.
John Crisanti

The people who made it happen: UP's steam crew poses for a portrait on the front of 4014 in Cheyenne the day prior to departure for Ogden. *Steve Glischinski*

Cylinder packing gets adjustments on May 3 after the test run. *Jim Wrinn*

Work continued on 4014 almost up to departure. A contractor works to attach banding to 4014's boiler jacket on May 3. *Three photos: Steve Glischinski*

UP 4-8-4 No. 844 is fired up in the Cheyenne Steam Shop on May 3.
Steve Glischinski

No. 844 gets a last minute bath as the crew puts a fire in the engine on the eve of the westbound journey.

"Last of the Giants" was the title of UP's Big Boy promotional film from the 1950s, and UP's last two operating giants share the floor of the steam shop the day before departure for Ogden. *Three photos: Steve Glischinski*

Inside the Cheyenne steam shop, workers ready UP 4014 for its big debut.

Big Boy 4014 shows off its long boiler inside the confines of the steam shop on May 3.

Big Boy test run: It's after 6 p.m., on May 2, when Big Boy slips under the Colorado & Southern (now BNSF) bridge on the west side of the Cheyenne yard. In the 1950s, this was a classic location for photographing Big Boy locomotives as they headed west out of Cheyenne. *Jim Wrinn*

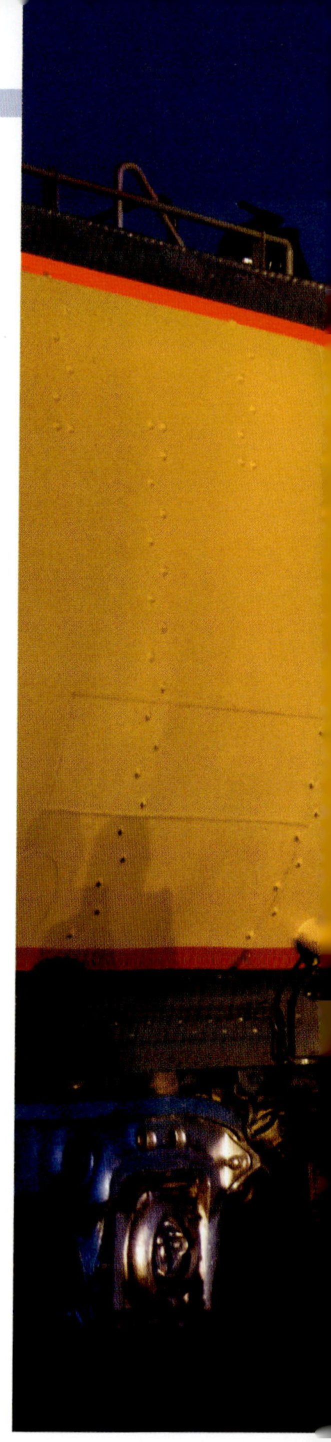

Well after sunset, No. 4014 stops at Speer Junction, just south of Cheyenne, on May 2, 2019. Illuminating the scene is a waiting manifest about to enter the Greeley Sub for the run south into Colorado. *John Crisanti*

Extra 4014 West

Big Boy's inaugural trip returns it to the rails that made it famous

"EVERYTHING IS POSSIBLE, EVEN THE IMPOSSIBLE" is a quote from the 2019 Disney remake of *Mary Poppins*. It is a line that was appropriate in Wyoming and Utah May 4-9, 2019, as Big Boy No. 4014 made its inaugural run from Cheyenne, Wyo., to Ogden, Utah, in celebration of the 150th anniversary of the Golden Spike.

With work on Big Boy at a frenzied pace leading up to the trip, 4-8-4 No. 844 did not make a planned solo run to Ogden the week before and instead doubleheaded with the newcomer. Their duet made for an amazing sight as the Big Boy and the Living Legend thrilled spectators.

Thousands of onlookers from across the country and around the world crowded roads and populated every vacant spot along the UP main line following a christening ceremony headlined by UP Chairman, President, and CEO Lance Fritz; Wyoming Gov. Mark Gordon; and Cheyenne Mayor Marian Orr. Fritz's wife Julie smashed a bottle of champagne on the pilot of the locomotive before a crowd of several hundred who paid to see the event at the Cheyenne depot.

Then it was off to climb Sherman Hill via Track 3, the locomotive returning to Big Boy territory for the first time since July 1959. The train overnighted in three towns along the way Rawlins, Rock Springs, and Evanston — before a triumphant return to Wasatch grade. The two left Rock Springs early on the morning of May 6 to get ahead of track work, and they continuously dodged freights on the UP's busy Overland Route.

In Ogden, Nos. 4014 and 844 participated in a special ceremony in which they appeared nose-to-nose in the same fashion as the Central Pacific 4-4-0 Jupiter and Union Pacific 4-4-0 No. 119 in 1869. With a call on the radio from UP's Fritz as part of a Thursday morning ceremony, Big Boy No. 4014 eased into the scene with No. 844 in front of a banner that read "#Done." Fritz, Utah Gov. Gary Herbert, U.S. Rep. Bob Bishop, and descendants of UP construction boss Grenville Dodge and Chinese laborers used hammers to tap an oversized golden spike, celebrating the 150th anniversary of the first transcontinental railroad in 1869. Big Boy had fulfilled its aim of being the biggest gift to the transcontinental railroad's 150th anniversary.

Locomotives 4014 and 844 lead their special train headed to the Golden Spike 150 Celebration in Ogden at Sinclair, Wyo., on May 4, 2019. *Mike Danneman*

Across Sherman Hill for the first time in 60 years

> Dignitaries crowd the platform at the Cheyenne depot on May 4 as Nos. 4014 and 844 hold the westbound main line. In a few minutes, No. 4014 will become the first Big Boy in steam on its original route in 60 years. *Jim Wrinn*

∧ UP President Lance Fritz and his wife Julie break a bottle of champagne to christen the engine on May 4, 2019. After the speeches were over, the two locomotives eased ahead to pick up passengers and head across Sherman Hill to Laramie and eventually Rawlins. *Jim Wrinn*

Engineer Ed Dickens talks to admirers from the gangway of Union Pacific Big Boy No. 4014 at Rawlins, Wyo., on May 4, 2019. The script on the side of the cab has changed: The fractions are gone from the cylinder dimensions, as are two letters that once described the stoker type when the engine burned coal. In their place are "DB" for Dickens-Barker, the two steam shop crew members who installed the oil burner. *Mike Danneman*

Across Sherman Hill for the first time in 60 years

The 4014 and 844 roll westbound near Harriman, Wyo., on Track 3, a Sherman Hill alignment that only saw Big Boys from 1953 to 1959. *Steve Glischinski*

< **Number 4014 pounds up Sherman Hill Track 3 east of Harriman, Wyo.** *Steve Glischinski*

< **The duo join tracks 1 and 2 at Dale Junction, Wyo., as a westbound freight waits for them to pass before following the special through the Hermosa Tunnels on May 4, 2019.** *Jim Wrinn*

Across Sherman Hill for the first time in 60 years

^ Steam crew members lube No. 4014 at Rock Springs, Wyo. *Mike Danneman*

V Big Boy and No. 844 make a spirited departure from Point of Rocks, Wyo., after a servicing stop on May 4, 2019. *Robert S. McGonigal*

> Number 4014 highballs at some 35 mph between Rawlins and Wamsutter, Wyo. The medallion on the floor covers a drawbar pin inspection opening. *Jim Wrinn*

Across Wyoming

With an early morning departure from Rock Springs, Wyo., on May 6, 2019, confounding many photographers, some chose to chase the sun. Those who were at the junction town of Granger, Wyo., were richly rewarded with a dazzling sunrise and doubleheaded UP steam locomotives in their stride.
Chase Gunnoe

X4014

Back to Echo Canyon

> On May 8, 2019, Nos. 4104 and 844 pop out of the tunnel at Curvo, Utah, on the 1869 line. Below at this famous spot is the 1916 second track for eastbound trains.
Two photos, Jim Wrinn

No. 4014 steams through Castle Rock, Utah, on a cold, rainy morning on May 8, 2019. *Mike Danneman*

The 4014's westbound passenger car special, symboled PENOG1-08, rides the Evanston Subdivision at milepost 935. It's descending Echo Canyon at Castle Rock, Utah. The train is on the original 1869 grade. On the hillside above is the 1916 second track. *James Belmont*

Nos. 4014 and 844 sprint on the last lap at Henefer, Utah, before reaching Ogden and their objective. Following along on parallel I-84 is a motorcade that by most observers' perspectives reached 12 miles in length. *Jim Wrinn*

Celebrating the Golden Spike in Ogden, Utah

> **Standing in for the *Jupiter* and No. 119 in a remake of the classic East-meets-West Golden Spike pose, the 4014 and 844 face each other at Ogden, Utah, on May 9, 2019, to celebrate the 150th anniversary of the Golden Spike.** *Two photos, Jim Wrinn*

Union Pacific officials and guests reach for hammers to ceremonially tap a giant Golden Spike, marking the completion of the first transcontinental railroad on May 10, 1869.

Extra 4014 East

Big Boy attacks its old foe, Wasatch Grade, on its way home to Cheyenne

AFTER THE BIG PARTY IN OGDEN, it was time for Nos. 4014 and 844 to return home to Cheyenne, retracing their 500-mile route. The first challenge: the 65-mile-long Wasatch grade for which the Big Boy locomotive was designed. The grade starts in Ogden and ends at Wahsatch, Utah, just east of Evanston, Wyo. While No. 4014's May 12 passenger special was nowhere near the tonnage it once pulled in the freight era, it was no less impressive in scenic Weber and Echo canyons.

The train overnighted in Evanston and paused in Rock Springs for two days to rest the crew and provide visitors the chance to inspect the engine more closely.

On May 16, the special ran to Rawlins on an uneventful trip until entering the yard, where axles 2 and 3 of No. 4014's second engine dropped between the gauge. Fortunately the train was running slowly and the crew was attentive and stopped the engine quickly. With the help of wood blocking and a local track crew, the engine was rerailed and on its way in 3 hours.

The following day, No. 4014 worked its way through heavy freight traffic and a tie gang near Walcott, Wyo., and ended up in Laramie for another day of rest.

The grand finale was the eastbound trip across legendary Sherman Hill. The engine pierced the Hermosa Tunnels and dropped downhill on tracks 1 and 2 for the first time. At Cheyenne, the two steam locomotives cut off from their train and headed to the shop for a well-deserved rest with more than 1,000 miles, a tremendous amount of hard work, and millions of smiles and memories behind them.

This article appeared in Big Boy: Back in Steam, *a* Trains *special edition, in 2019.*

In the heart of Echo Canyon, Nos. 4014 and 844 climb fabled Wasatch Grade, the operating obstacle the Big Boy was designed to challenge in 1941. *James Belmont*

X4014
4014
UNION PACIFIC

Climbing Wasatch Grade

With a long motorcade hot in pursuit on parallel Interstate 84, Nos. 4014 and 844 thread a bridge over Weber Creek near Morgan, Utah, on May 12, 2019.
Sol Tucker

UNION PACIFIC

Looking like a model locomotive on a train layout, No. 4014 darts out of a tunnel in Weber Canyon.
Sol Tucker

With the abandoned summit tunnel to the left, the doubleheader reaches the summit at Wahsatch, Utah, on the 1916 line (the town has an "h"; the mountains do not). The original 1869 grade is on the top of the embankment over the locomotives. The sound here: like that of a jet airliner taking off. *Jim Wrinn*

X4014
4014

Grand send-off

Wyoming is cowboy and cowgirl country, and that was evident in the send-off Nos. 4014 and 844 received on the morning of May 13 as they departed Evanston, Wyo.
Leo de Groot

The 1929 welcome sign from the Lincoln Highway, restored and repositioned at the subway under the UP tracks, frames the 4014's departure from Rock Springs, Wyo., on May 16, 2019. *Jim Wrinn*

No. 4014's departure from Rock Springs, Wyo.

> **Employee tags hang in the cab.**
John Crisanti

V **Ed Dickens checks the fire.**
John Crisanti

No. 4014 pauses near the steam-era water tank at Wamsutter, Wyo.
John Crisanti

When the drivers on axles 2 and 3 on the second engine dropped between the rails in Rawlins yard, the crew was able to block the wheels and re-rail the locomotive in short order.
Jim Wrinn

Return to Laramie

> Rain showers move in as Nos. 4014 and 844 put on a show on departure from Rawlins, Wyo., on May 17, 2019.
John Crisanti

The famous rock formation east of Rock River, Wyo., where the main line climbs a ridge, has been a favorite of photographers on the UP since the steam era. It was no less so on May 17, 2019, as No. 4014 made its first eastbound trek under steam. *Jim Wrinn*

Back home to Cheyenne

Bathed in a cloud of steam and smoke, No. 4014 pops out of Hermosa Tunnel on Sherman Hill on May 19, 2019. *James Hickey*

The 12-car special, with office car St. Louis bringing up the tail, rolls downgrade at Granite on the last few minutes into Cheyenne. *Jim Wrinn*

The two locomotives cut off their train on the main line and proceed to the steam shop in Cheyenne.
Two photos, Jim Wrinn

With No. 844 hostling the auxiliary tenders, No. 4014 takes its rest outside of the steam shop at the end of the tour on May 19, 2019.

9

WHERE THE BOYS ARE

Along with 4014, seven other Big Boys have been preserved

by Jim Wrinn

Eight Big Boys have been preserved, a remarkable number considering only 25 of the class were built. Here No. 4006 gets adjustments in Kansas City, Mo., on the Missouri Pacific on its way to the Museum of Transportation in St. Louis in 1961. *Ron Christison*

NO. 4004
Holliday Park, Cheyenne, Wyo.

BUILDER'S DATE: September 1941

SERIAL NUMBER: 69575

HISTORY: The fifth Big Boy constructed seems to have had an uneventful career.

LIFETIME MILEAGE: 1,060,402

DATE RETIRED: February 1962

PRESERVATION: According to Bess Arnold's 2004 book, *Union Pacific: Saving a Big Boy and other railroad stories*, the 4004's salvation is the direct result of a retirees group that recognized how quickly the railroad was disposing of its steam power. The group reached out to UP Vice President-Operations E.H. Bailey about saving a Big Boy. Bailey concurred, and brakeman Fred Mueller, Dr. Ralph Gramlich, and retired machinist Frank Hardy went to the mayor's office in Cheyenne, Wyo., where they got approval to preserve an engine. The UP Old Timers Club sold raffle tickets and collected more than $2,000 to build a concrete pad for the locomotive in what was once Holliday Park's Lake Minnehaha. A diesel switcher moved the locomotive across 600 feet of panel track on June 28, 1963. As an extra precaution, a city truck followed to feed air to the locomotive's brake system in case it broke free from the diesel, and a crawler tractor was tethered to the 4004 as a further safeguard to prevent it from rolling free. The locomotive was flooded in 1986 when heavy rains caused ponding in the park and enough water accumulated to almost cover the 68-inch drivers. Small parts have been removed for use on UP's active steam engines.

CURRENT STATUS: On display in Holliday Park, illuminated, and accessible 24 hours a day. The tender was delivered with sister Big Boy No. 4002.

Number 4004's builder's plate is chipped in the corner. *Jim Wrinn*

With clouds overhead mimicking steam, Big Boy No. 4004 sleeps in October 2013 at Cheyenne's Holliday Park, just a few blocks from UP rails. *Jim Wrinn*

Tucked inside the Forney Transportation Museum in Denver, No. 4005 tells the Big Boy story up close. *Jim Wrinn*

NO. 4005

Forney Museum of Transportation, Denver

BUILDER'S DATE: September 1941

SERIAL NUMBER: 69576

HISTORY: Of all the preserved Big Boys, No. 4005 has more of a story than the others. No. 4005 was the only Big Boy converted to burn oil as an experiment in 1946, during a miner's strike. The locomotive operated in this fashion through March 1948, but the test was deemed unsuccessful and the engine was reconfigured to burn coal, which it did for the rest of its operating life. The locomotive was involved in a fatal derailment on April 27, 1953, on a run from Rawlins, Wyo., to Green River, Wyo. It was pulling 62 cars and a caboose when it entered an open siding switch at Red Desert and derailed at 50 mph. The locomotive and tender came to rest on their left sides and the first 18 cars piled up in the crash, which killed the engineer, fireman, and head-end brakeman. Number 4005 was repaired at Cheyenne and returned to service, still bearing scars of the accident on the left side.

LIFETIME MILEAGE: 1,043,624

DATE RETIRED: July 1962

PRESERVATION: Donated to the museum in downtown Denver in 1970. The museum was relocated to its present location in 1998 and reopened in 2001.

CURRENT STATUS: On display with other rolling stock inside the Forney Museum of Transportation, 4303 Brighton Blvd., Denver, Colo. The cab is open to visitors and the backhead is complete with parts labeled. Visitors wishing to get good photos should bring the widest-angle lens possible, owing to the cramped location of the locomotive. For more information about seeing No. 4005, visit www.forneymuseum.org.

The 4005 is displayed with a complete backhead, and all of the parts and controls are labeled. *Jim Wrinn*

Here's a most unlikely meeting in June 1962 between No. 4006 and the famous Civil War locomotive *General*, an 1855 4-4-0 that was touring the country.
Museum of Transportation

NO. 4006
Museum of Transportation, St. Louis

BUILDER'S DATE: September 1941

SERIAL NUMBER: 69577

LIFETIME MILEAGE: 1,064,625. This was the highest-mileage Big Boy of the 25 locomotives in this class. The tender is from No. 4003.

DATE RETIRED: May 1961

PRESERVATION: Negotiations to save a Big Boy for the Museum of Transport (as it was known then) began in 1954, long before the 4-8-8-4s were withdrawn from service. UP formally donated the locomotive to the museum in June 1961, moving it to Kansas City, Mo., in a journey that took four days at a top speed of 25 mph. In Kansas City, Missouri Pacific took over for the rest of the trip, but instead of going straight to the museum, the engine went to the Alton & Southern shop in East St. Louis for cosmetic work that took about a year. During its delivery to the museum, the locomotive encountered one of the best-known steam locomotives in American history, the American-type *General* of Civil War "Great Locomotive Chase" fame. The meeting took place when No. 4006 left the Alton & Southern yard in East St. Louis on June 5, 1962. With two diesels pulling it, No. 4006 rolled across the MacArthur Bridge over the Mississippi River. The *General* was in St. Louis on the first leg of a Civil War centennial tour. Said the Museum of Transport in an announcement, "One hundred years of steam locomotive development were embraced in the encounter between the 600-ton articulated giant built in 1941 and the 31-ton 4-4-0 constructed in 1855."

CURRENT STATUS: Recently repainted using more than 45 gallons of finish coat. For more information on the locomotive, visit the museum's Web page at transportmuseumassociation.org.

This January 2014 image of No. 4006 shows the significant work that took place to restore this mammoth locomotive to the appearance of an operating Big Boy. *Mark Mautner*

NO. 4012

Steamtown National Historic Site, Scranton, Pa.

BUILDER'S DATE: November 1941

SERIAL NUMBER: 69583

MILEAGE: 1,029,507

DATE RETIRED: February 1962

PRESERVATION: The 4012's 1964 trip from the UP yard at Council Bluffs, Iowa, to Bellows Falls, Vt., to become the largest and most impressive piece in Nelson Blount's extensive Steamtown U.S.A. collection, was not without excitement. First off, according to James R. Adair's 1967 book, "The Man from Steamtown," it cost Blount $6,000 just to ship the locomotive dead-in-tow. Then, the centipede tender derailed three wheels at Manchester, N.Y., causing much consternation among the crews who had to rerail the giant tank.

CURRENT STATUS: The locomotive has been displayed outside since the mid-1960s. Relocated to Scranton, Pa., with the rest of the Steamtown collection in 1984, it is one of the major attractions at Steamtown National Historic Site, an operation of the National Park Service. For more information about visiting No. 4012, go to www.nps.gov/stea/.

Inset: In fall 1964, No. 4012 was moving at Mechanicville, N.Y., with a sign on its boiler that read (incorrectly boasting of its size), "I'm the largest loco in the world, going to Steamtown, Bellows Falls, Vt." *Jim Shaughnessy*

Big Boy 4012 stands near the entrance of the Steamtown National Historic Site in Scranton, Pa., where Delaware, Lackawanna & Western 4-8-4s once held sway.
Jim Wrinn

Big Boy No. 4017 resides indoors inside the National Railroad Museum's Lenfestey Center alongside a famous Pennsylvania Railroad GG1 electric locomotive. The 4017 is one of the most complete Big Boys on display. *Brian Schmidt*

NO. 4017
National Railroad Museum, Green Bay, Wis.

BUILDER'S DATE: December 1941

SERIAL NUMBER: 69588

LIFETIME MILEAGE: 1,052,072

DATE RETIRED: May 1961

PRESERVATION: Letters flew in 1960 and 1961 between the museum's Harold Fuller and UP leadership, including Vice President-Operations E.H. Bailey, President A.E. Stoddard, and board Chairman E. Roland Harriman. The letters introduced the young museum and laid the groundwork for the Big Boy donation. Once accomplished, the next issue was getting it from Cheyenne to Green Bay. The Chicago & North Western was set to take No. 4017 across Iowa, Illinois, and into Wisconsin. At the last minute, the Milwaukee Road stepped in and asked for the honor of moving the engine. The Milwaukee Road did so, but couldn't claim all the glory: The Milwaukee Road still had to hand No. 4017 over to the C&NW for final delivery.

CURRENT STATUS: On display inside the climate-controlled Lenfestey Center, No. 4017 is the centerpiece of the exhibit hall alongside North America's most popular electric locomotive, a Pennsylvania Railroad GG1. To learn more, visit the National Railroad Museum's website at www.nationalrrmuseum.org.

This is the view inside No. 4017's firebox, looking in from the firedoor toward the back tube sheet. Circulators that allow water to move inside the boiler form inverted Ts. *Brian Schmidt*

Number 4017's tender comes from No. 4023, and is the only surviving tender from from the 1944 series of Big Boys. The coal stoker auger is visible at bottom. High-sloped side sheets allow for an additional 1,000 gallons of water compared to tenders from the original order. *Jim Wrinn*

NO. 4018
Museum of the American Railroad, Frisco, Texas

BUILDER'S DATE: December 1941

SERIAL NUMBER: 69589

LIFETIME MILEAGE: 1,037,123

DATE RETIRED: July 1962

HISTORY: No. 4018 was shopped at Cheyenne in April 1957 and ran the following September. This was short-lived, and it was stored at Green River by October 1958, never to run again.

PRESERVATION: The locomotive arrived at the Age of Steam museum at the Texas State Fair in Dallas in 1964. In 1998 the museum was approached with plans to restore 4018 to operation for a movie. This plan never materialized.

CURRENT STATUS: The locomotive and the museum relocated in 2013. Scott Lindsay coordinated No. 4018's move and says: "The complexity of the overall design of the 4000 class is an engineering marvel. It makes you appreciate the skill of the crafts that maintained these locomotives. That said, the Big Boys were not designed to be serviced without a shop and pit. Fortunately we had great museum volunteers, including some who were thin enough to climb under to drain and refill the driver roller bearings. Once we were ready to move No. 4018 to the lead for testing, I was impressed how the locomotive easily negotiated the tight curves within the museum. The tender proved to have typical difficulty with the curves even after jacking all but one of the rigid-wheelbase wheel sets above the rail. We ensured that all crossheads, the bearing plate for the front engine, shoes, wedges, rods, and other critical surfaces received lubrication. The thick cab insulation, which kept crews warm in cold Wyoming winters, works well to overheat crews moving one at low speed in the Texas summer heat." For more details about this Big Boy, see www.museumoftheamericanrailroad.org.

Complete, save for the main rods, Big Boy No. 4018 looks good in the Texas sun prior to its trip from the Dallas fairgrounds to its new, permanent home in Frisco. *John Hewitt*

Number 4018's giant centipede tender fills the sky in this view after its trip over BNSF Railway tracks to its new display site in Frisco. The locomotive carries its original tender. *Ken Fitzgerald*

From its perch on a hillside in Lauritzen Gardens, No. 4023 overlooks the Missouri River and Iowa to the east as well as Union Pacific's Omaha yard and Interstate 80 below. *Two photos: Cate Kratville*

NO. 4023
Lauritzen Gardens, Omaha, Neb.

BUILDER'S DATE: November 1944

SERIAL NUMBER: 72781

HISTORY: One of five Big Boy locomotives built to supplement the original 20.

LIFETIME MILEAGE: 829,295

DATE RETIRED: July 1962

PRESERVATION: Big Boy No. 4023 was kept inside the roundhouse at Cheyenne, Wyo. (with Challenger No. 3985) until the mid-1970s when it was relocated to Omaha and placed adjacent to the UP shops there. A steam exhaust line was fashioned so that it appeared that smoke and steam were coming from the twin stacks. After the shops closed, the engine was relocated to the Durham Western History Museum in downtown Omaha and repositioned in 2004 to its current location at Kenefick Park in Lauritzen Gardens. This move required a trip over city streets in a special cradle designed to spread the weight of the locomotive. Wasatch Railroad Contractors performed a complete cosmetic restoration of the steam locomotive during the five months following the move. In addition to a new jacket, many functional appliances were replaced with new, mock appliances. This included the safety valves, whistle, and lubricators.

CURRENT STATUS: The only surviving example of the last class of five Big Boys welcomes visitors along Interstate 80 to UP's headquarters city. The locomotive is perched on a hillside overlooking the Missouri River in the company of a UP DDA40X diesel locomotive, one of the largest diesels ever built. Details: www.lauritzengardens.org/about/kenefick_park.

The original Centennial DDA40X diesel locomotive and the next-to-last Big Boy share a place of honor in Kenefick Park in Omaha in a salute to John Kenefick, the longtime Union Pacific leader in the 1970s and 1980s.

Number 4018 pushes an eastbound manifest freight up Sherman Hill's western slope at Forelle, Wyo., on Nov. 15, 1954. Eastbounds tackled the hill on the "new line," built in 1901, while westbounds sailed down the 1868 main line. *Union Pacific; Al Chione collection*

BIG BOY GALLERY

A look at 4-8-8-4s in action and at rest

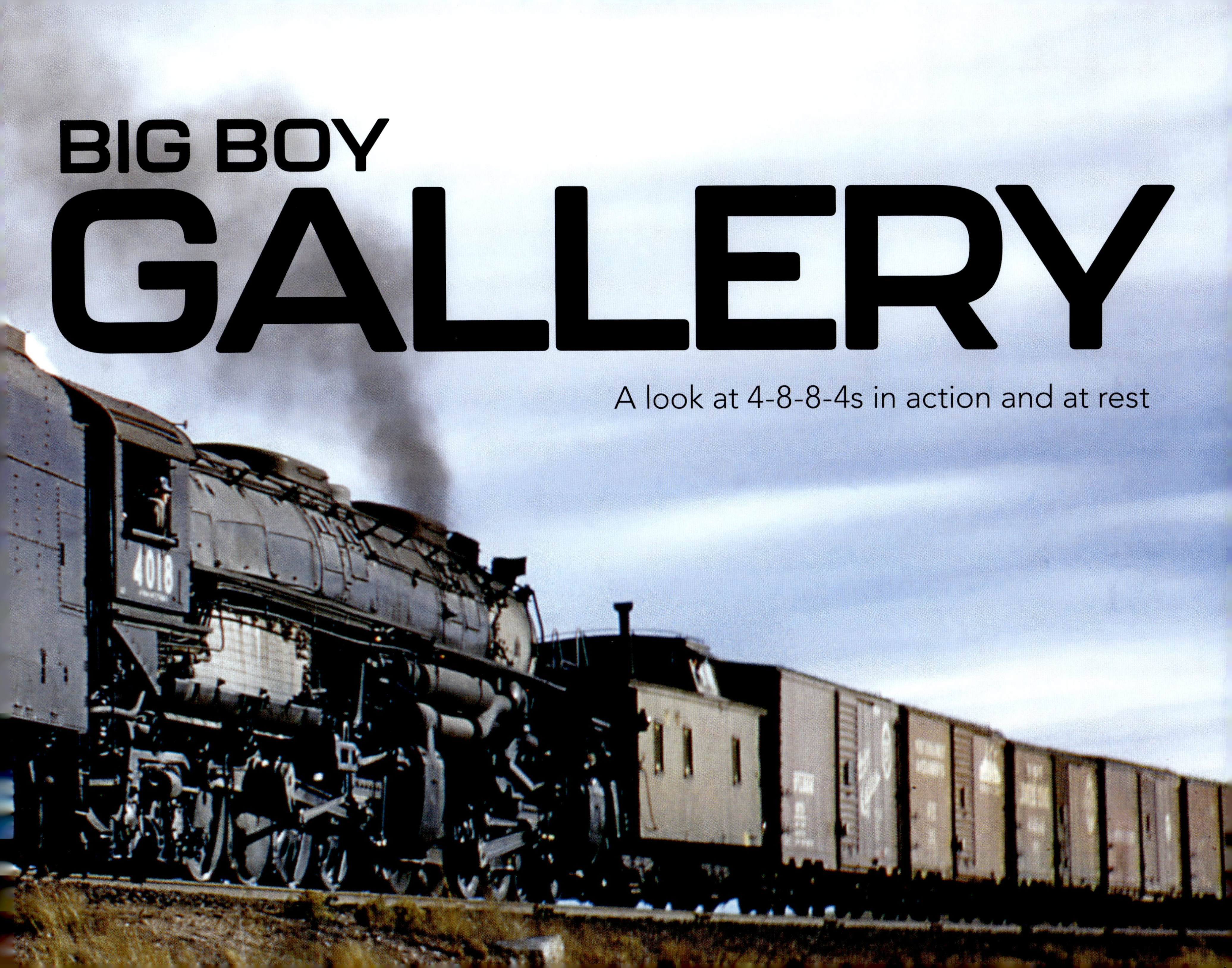

TRAINS, AND STEAM LOCOMOTIVES IN PARTICULAR, are especially photogenic. Because of their size, the contrast they strike with the landscape, and their forward motion, they are fleeting glimpses of mankind's incredible ingenuity and mechanical splendor. Sitting still, they lend themselves to portraiture. On the road, they are the prize in action photography. Big Boy locomotives were especially sought after by enthusiastic railroad photographers far and wide. They were the big game of the railroad.

In the day before jet airliners or Interstate highways, photographers made pilgrimages down gravel and dirt roads to Cheyenne, Green River, and Ogden. They spent lonely hours trackside in Echo Canyon, Peru Hill, and Sherman Hill. We are fortunate that they trekked into roundhouses, backshops, and servicing facilities so we all can see these massive beasts as they were being groomed and fed for work. We are blessed that they stalked their prey at desolate spots like Dale Junction, Castle Rock, and Bitter Creek. The photographic record of Big Boys in regular service was made in 18 short years in the middle of the 20th century.

Because of images likes the ones on these pages, these magnificent engines live on as they once were — giants of the rails among us.

The UP had to upgrade turntables at several locations to accommodate the Big Boys. Number 4023 takes up almost every inch of the 126-foot table at Cheyenne. *E.C. Storm*

Big Boy 4017 roars out of Cheyenne, Wyo., in the teeth of an April 1, 1957 blizzard that had shut down road traffic. Yet UP hadn't even called out a plow, knowing that 14 inches of snow is no match for the biggest power on the line.
Frank Barry

Blasting out of the 1,800-foot Hermosa Tunnels, Big Boy 4010 claws its way eastbound up Wyoming's Sherman Hill in 1956.
Leroy Dunn; James L. Ehernberger collection

4010
UNION PACIFIC

While hurrying a long freight east of Medicine Bow, Wyo., a Big Boy meets its future, and its fate, in the rounded nose of a diesel headed the opposite direction. *Robert Hale*

The giant tender of a 4-8-8-4 eases through a wash rack, rinsing away the layers of grime and soot accumulated with every run across Wyoming. *Robert Hale*

Framed by the cylinder of Big Boy No. 4022, a mechanic inside the Cheyenne, Wyo., roundhouse checks the locomotive work schedule board in 1957. *Philip R. Hastings; Kalmbach Books collection*

Big Boy 4022 has a full head of steam leaving Cheyenne, Wyo., in 1958. Gunning for Sherman Hill, the engine blasts under the Colorado & Southern bridge with a westbound freight. Billboards along the Lincoln Highway appear at left. *Stan Kistler*

Number 4016 storms east with 103 cars (including 15 stock cars behind the tender) at Green River, Wyo., on Oct. 12, 1957. Moving long, heavy trains great distances: This was Big Boy's legacy. *Jack Pfeifer; Al Chione collection*

On the man-made horizon at Dale, Wyo., Big Boy No. 4019 teams up with an A-B-A set of EMD F3 diesels to forward tonnage west on Sept. 12, 1956. *Dave Rorick*

The biggest in steam slakes its thirst as a water plug dipped inside a Big Boy's tender sends thousands of gallons cascading in. That tender can hold 24,000 gallons of water. *Robert Hale*